THE INFORMATION REVOLUTION IN MILITARY AFFAIRS IN ASIA

THE INFORMATION REVOLUTION IN MILITARY AFFAIRS IN ASIA

Edited by

Emily O. Goldman

and

Thomas G. Mahnken

THE INFORMATION REVOLUTION IN MILITARY AFFAIRS
© Emily Goldman and Thomas Mahnken, 2004

First published 2004 by
PALGRAVE MACMILLAN™
175 Fifth Avenue, New York, N.Y. 10010 and
Houndmills, Basingstoke, Hampshire, England RG21 6XS
Companies and representatives throughout the world

PALGRAVE MACMILLAN is the global academic imprint of the Palgrave Macmillan division of St. Martin's Press, LLC and of Palgrave Macmillan Ltd. Macmillan® is a registered trademark in the United States, United Kingdom and other countries. Palgrave is a registered trademark in the European Union and other countries.

ISBN 1–4039–6467–X hardback   ISBN: 978-14039-6467-0

Library of Congress Cataloging-in-Publication Data
     The information revolution in military affairs : the prospects for Asia / edited by Emily Goldman and Thomas Mahnken.
        p. cm.
     Includes bibliographical references and index.
     ISBN 1–4039–6467–X
        1. Information warfare—East Asia. 2. Information warfare-Australia.
     3. Military art and science-East Asia—History-21st century. 4. Military art and science-Australia—History-21st century. 5. East Asia—Defenses.
     6. Australia—Defenses. 7. World politics—1995–2005 I. Goldman, Emily O., 1961– II. Mahnken, Thomas G., 1965–

UA832.5.I54 2004
355'.03305—dc22   2004044254

A catalogue record for this book is available from the British Library.

Design by Newgen Imaging Systems (P) Ltd., Chennai, India.

First edition: July 2004
10 9 8 7 6 5 4 3 2 1

Transferred to Digital Printing 2011

TABLE OF CONTENTS

ACKNOWLEDGMENTS

This book would not have come about without the support of Mr. Andrew W. Marshall, Director of Net Assessment in the U.S. Department of Defense.

In addition, a number of people helped make this book a reality, including Steve Ayling, Ross Babbage, Desmond Ball, Paul Dibb, James R. FitzSimonds, and Narushige Michishita. We would also like to thank our research assistants: Ani Ahn, Leo Blanken, Satoshi Shimada, and John Kennedy.

We would also like to thank our editor at Palgrave Macmillan, Mr. Toby Wahl, as well as his assistant, Heather van Dusen, for their enthusiastic support of this project.

Notes on Contributors

Adam Cobb is the founder and Director of Stratwise, Australia's first privately funded national security think tank. He has held a range of international academic and government positions, including Special Director—Strategic Policy in Air Force Headquarters, Senior Defence Adviser in the Federal Parliament, the University of Amsterdam, the Strategic and Defence Studies Centre (ANU), and Sydney University. Cobb has also worked in the Australian Embassy in Washington DC and for a member of the U.S. House of Representatives. Considered an authority on asymmetric warfare, his analysis is regularly sought by the BBC, ABC, commercial TV News, newspapers and radio. Most recently, he was commissioned by SBS World News (broadcast nationally in Australia) as a military analyst providing commentary and analysis of the 2003 Gulf War for the nightly news.

Arthur S. Ding is a Research Fellow at the Research Division III (China Politics) of the Institute of International Relations at National Chengchi University in Taipei, Taiwan. His research focuses on China security and defense issues. He has published *China's Military Theory in Change: 1979–1991* and *Defense Industry Conversion in China*, as well as numerous articles. He is currently a Fulbright Scholar associated with Harvard University's Fairbank Center for East Asian Research. He obtained his Ph.D. in Government and International Studies from the University of Notre Dame in 1987.

Michael Evans is Head of the Australian Army's Land Warfare Studies Centre (LWSC) at the Royal Military College, Duntroon in Canberra. He has served on the staff of Australian Land Headquarters and in the Australian Directorate of Army Research and Analysis. Dr. Evans is a graduate in history and war studies of the universities of Rhodesia (B.A. Hons), London (M.A.) and Western Australia (Ph.D.). He has been a Sir Alfred Beit Fellow in the Department of War Studies at King's College, University of London and a Visiting Fellow at the University of York in England. He recently co-edited (with Alan Ryan) a major study of future war, *Future Armies, Future Challenges: Land Warfare in the Information Age* (Allen & Unwin, Sydney, 2003). He is currently writing a book entitled, *Military Theory and 21st Century Warfare: The Legacy of the Past and the Challenge of the Future*.

Emily O. Goldman is Associate Professor of Political Science at the University of California, Davis, and Director of the UC Davis Washington

Program. She was a residential fellow at the Woodrow Wilson International Center for Scholars during 2003–04. She also co-directs the Joint Center for International and Security Studies (JCISS), a research and educational partnership between UC Davis and the Naval Postgraduate School in Monterey. Her publications include *Sunken Treaties: Naval Arms Control between the Wars* (Penn State Press, 1994); *Diffusion of Military Technology and Ideas*, with Leslie C. Eliason, eds. (Stanford University Press, 2003); *The Politics of Strategic Adjustment*, with Peter Trubowitz and Edward Rhodes, eds. (Columbia University Press, 1999).

Tim Huxley is Senior Fellow for Asia-Pacific Security as well as Editor, Adelphi Papers, at the International Institute for Strategic Studies in London. He was previously at the University of Hull, where he was Reader in Southeast Asian Politics and Director of the Centre for Southeast Asian Studies. Educated at Oxford (MA), Aberystwyth (MScEcon) and the Australian National University (PhD), he has held appointments at the University of New South Wales, the Institute of Southeast Asian Studies (Singapore), Lancaster University, the University of Wales Aberystwyth, and in the Australian Parliamentary Research Service. His research interests focus on Southeast Asian defense and security issues. Among other publications, he is the author of *Defending the Lion City. The Armed Forces of Singapore* (Allen & Unwin, 2000) and two IISS Adelphi Papers: *Arming East Asia* (No 329, coauthored with Susan Willett) and *Disintegrating Indonesia? Implications for Regional Security* (No 349).

You Ji is Senior Lecturer in School of Political Science, the University of New South Wales. He has published widely on China's political, economic, military, and foreign affairs. He is author of three books: *In Quest of High Tech Power: the Modernization of China's Military in the 1990s* (1996); *China's Enterprise Reform: Changing State/Society Relations after Mao* (1998); and *The Armed Forces of China* (1999). His most recent articles include: "Profile: the Heir Apparent," *China Journal*, Vol. 49, July 2002; "The Supreme Leader and the Military," in Jonathan Unger (ed.) *The Nature of Chinese Politics: from Mao to Jiang*, Armonk: ME Sharp, 2002.

Thomas G. Mahnken is a Professor of Strategy at the U.S. Naval War College. In 2003–04 he was Visiting Professor and Acting Director of the Strategic Studies Program at the Johns Hopkins University's Paul H. Nitze School of Advanced International Studies. He served in the Defense Department's Office of Net Assessment and as a member of the Gulf War Air Power Survey, commissioned by the Secretary of the Air Force to examine the performance of U.S. forces during the war with Iraq. Prior to that, he served as an analyst in the Non-Proliferation Directorate of the Office of the Secretary of Defense (OSD), where he was responsible for enforcing U.S. missile proliferation policy. He is the author of *Uncovering Ways of War: U.S. Intelligence and Foreign Military Innovation, 1918–1941* (Cornell University

Press, 2002); with James R. FitzSimonds of *The Limits of Transformation: Officer Attitudes toward the Revolution in Military Affairs* (Naval War College Press, 2003); and with Richard K. Betts editor of *Paradoxes of Strategic Intelligence* (Frank Cass, 2003). He is also the co-editor of the *Journal of Strategic Studies*.

James Mulvenon is a Political Scientist at the RAND Corporation in Washington, DC and Deputy Director of RAND's Center for Asia-Pacific Policy. A specialist on the Chinese military, his current research focuses on Chinese C^4ISR, strategic weapons doctrines (information warfare and nuclear warfare), theater ballistic missile defenses (TBMD) in Asia, Chinese military commercial divestiture, and the military and civilian implications of the information revolution in China. He has recently completed a book on the Chinese military's business empire, entitled *Soldiers of Fortune* (Armonk, NY: M.E. Sharpe, 2001). Dr. Mulvenon received his Ph.D. in political science from the University of California, Los Angeles.

Sugio Takahashi has been a Research Fellow of the National Institute for Defense Studies in Tokyo, Japan since 1997. He also served on the research staff of the Office of Strategic Studies, Japan Defense Agency from 1998–2001. His fields of study include conventional and nuclear strategy and Japan's defense policy. His publications include "The Impact of RMA on the Security of Japan," *The Journal of National Defense*, Vol. 27, No. 4 (December 2000) and "The Impact of BMD on the Defense Posture of Japan," *Ballistic Missile Defense and Security of Japan* (Tokyo: Japan Institute for International Affairs, 2003) (with Satoshi Morimoto).

Andrew Nien-Dzu Yang has worked at the National Sun Yat-sen University in Taipei, Taiwan since 1986. Since 1991 he has served as secretary general of the Chinese Council of Advanced Policy Studies (CAPS), a nonprofit Independent think tank in Taipei. He was trained as a specialist in security studies, particularly in the area of military competition and military balance in East Asia, the People's Liberation Army's modernization and its impact on East Asia security.

INTRODUCTION: MILITARY
DIFFUSION AND TRANSFORMATION

Emily O. Goldman

Defense planners in the United States have been captured by the notion of a revolution in military affairs (RMA), leaping ahead to embrace emerging technologies and transform how militaries go about their business. As the U.S. military develops the doctrine and organizational structures to exploit emerging technologies, U.S. strategists must pay attention to how those innovations are being adopted and adapted elsewhere. This external dimension is critical for U.S. strategy because the diffusion of military innovation undermines military preeminence. Innovation leaders do not monopolize "their" innovations for long, and are frequently surpassed by followers.[1] Leadership in effecting a military revolution is no guarantee of victory. Military revolutions are typically rooted in nonmilitary processes and so impossible to steer and control. Finally, wholesale replication of the innovations may be unnecessary to challenge or degrade the leader's capabilities.

To a greater extent than in any previous generation, U.S. military leaders today are aware that they are in the midst of an unfolding RMA. Even as they try to prod the process along in the U.S. armed forces in the hope of prolonging American military preeminence, they must attend to the RMA's diffusion abroad for dynamics outside the United States will determine the future of the current RMA as much as, if not more than, developments inside the United States.

This study builds on theoretical and historical insights about military innovation and diffusion to examine how the current RMA is unfolding in Asia and what the implications will be for regional and global power. The term "innovation" refers to radical changes in organizational structure, resource allocation, doctrine, and strategy. It encompasses the process of adapting the institutions and practices of war making to changing technological opportunities and/or social and political developments. In the United States, this process is called "transformation." Because military organizations are comparative institutions that closely monitor one another, innovation and diffusion are inextricably linked. New technologies and ideas spread from the settings in which they were originally conceived and developed, yet at the same time, militaries grapple with the application to their particular setting, which may involve unique adaptations and innovations. The focus of this project is on both facets—the diffusion of new war-fighting approaches to several Asian militaries from abroad, chiefly from the United States, and the unique process of innovation occurring within those military establishments. China, Japan, Taiwan, Singapore, and Australia are states identified by scholars as having the greatest capacity to exploit the emerging RMA. Paul Dibb has argued that Australia and Japan have a high capacity to absorb the RMA, while China, Singapore, and Taiwan have a moderate capability.[2] David Roessner and Michael Salamone have argued that both Singapore and Taiwan have exhibited long-term commitment to the expansion of their high-technology capacity.[3]

This chapter provides an overview of the global strategic context within which military innovation is unfolding in Asia. It discusses the methodology guiding the study, reviews the theoretical and empirical literature on military innovation and diffusion, and concludes with a summary of the key RMA enablers and barriers that will be critical for shaping how rapidly and deeply innovation proceeds across the region.

GLOBAL STRATEGIC ENVIRONMENT

A series of events at the onset of the twenty-first century will be crucial in shaping attitudes toward the RMA in the United States and abroad. Operation *Enduring Freedom* in Afghanistan and Operation *Iraqi Freedom* in Iraq demonstrated the technological prowess of the U.S. military. In the Afghan campaign, the first military operation that displayed the capabilities of a transformed U.S. military, Special Operation Forces (SOF)-directed precision airpower was a crucial component of rapid success, particularly in combination with Northern Alliance ground troops.[4] The U.S. military succeeded in projecting power over vast distances with relatively small numbers of troops.

The war in Iraq provided a laboratory to test the "high-tech low numbers-on-the-ground" strategy of RMA proponents. Some have questioned whether the Iraq war was a strong test of U.S. capabilities since Iraq did not prove to be a formidable opponent. Others point out that operations in Afghanistan or Iraq are still ongoing, and although new technologies greatly assisted the United States in defeating foes on the ground, they have not secured the peace. Nonetheless, these engagements have spurred the military leaderships of states that might want to challenge the United States in the future to reconsider their military preparedness and defense postures. The Chinese learned a great deal from recent U.S. operations and are determined to transform the People's Liberation Army to the extent financially possible, as well as to reassess their military posture toward Taiwan.

U.S. prowess will also likely accelerate efforts of weaker adversaries to counter American capabilities. While some countermeasures are likely to be traditional—cover, concealment, dispersion, small-unit independent maneuver—other adversaries may opt for terrorism, weapons of mass destruction, or anti-access strategies to avoid taking the United States on in high intensity land warfare.[5] Countermeasures are also likely to include selective high-tech investment or mini-RMAs based on precision-guided weapons that are widely available on the international market. Just when U.S. transformation efforts may stall because American forces are tied down in Iraq, innovation efforts abroad may accelerate.

The global war on terror may also spur defense transformation. Terrorist attacks have declined in many parts of the world (like the Americas) and stabilized in others (like the Middle East), while in Asia they have been rising.[6] For a country like Australia that must make tough choices about how to spend scarce defense assets, transformation holds out the prospect of making their military forces more efficient and usable in defending far-flung interests and meeting a range of adversaries including low-tech challenges like terrorists and the failed states that breed them. To protect the infrastructure of modern democratic society in the Asian region, Australia believes it must accelerate its exploitation of the RMA. Australia is also determined to ensure its forces remain interoperable with those of the United States, to avoid irrelevance to the extent financially possible. If it succeeds in implementing its "middle power" approach to exploiting the RMA—sufficient modernization to ensure interoperability—Australia may become an attractive model for Japan.

Transformation in the United States and abroad will also be shaped by the proliferation of weapons of mass destruction and events on the Korean peninsula. RMA capabilities that support special operations forces can assist the United States and its allies in containing North Korea and stemming the spread of illegal arms exports without threatening a major conventional conflict that could be triggered with larger forces.[7] RMA capabilities increase the flexibility to launch preemptive strikes to prevent near nuclear states from acquiring operational capabilities. Information warfare assaults are also likely to be a crucial part of any conventional operations against North Korea. The situation in North Korea is likely to shape innovation in Asia in one additional and very crucial way. It may lower the political taboo against constitutional reform in Japan that prohibits development of nuclear weapons and a more offensive military posture.

STUDY OVERVIEW

China, Japan, Taiwan, Singapore, and Australia are all of interest to the United States and their military decisions will have a significant impact on security in Asia. They vary along several important theoretical dimensions: their relationship to the United States (from close ally to potential adversary); their motivations to adopt RMA innovations (from interoperability with the U.S. military to countering U.S. capabilities); and the factors that affect their capacity to integrate RMA innovations (from information-driven societies to minimally informatized societies and economies). These are a useful set of cases for mapping out RMA trajectories.

The research design follows the methodology of "diffusion diagnostics," which analyzes four key components of the diffusion process. First, one must identify the state's motives for importing and adopting new technologies, ideas, and practices. This requires understanding how national security establishments define the current RMA; how individual services view it; and how applicable the RMA is seen to be to the country's security situation.

Second, the models (regional or global) that are targets of emulation, adaptation, or offsets will provide important information on diffusion trajectories. The avenues of transnational communication and influence by which innovations are disseminated will also shape the course of diffusion. Who are defense officials communicating with in other countries about RMA issues? Is any particular military or set of militaries being held up as a model to be emulated? Whose concepts are being used or adapted? How much are developments within the United States influencing modernization efforts abroad?

Third, one must assess the capacity for military technology and ideas to be absorbed in different environments. What factors enable and constrain the spread of military knowledge? What factors hinder retention of the new idea by the receiving state and/or organization? What obstacles do RMA advocates face?

The final task is to capture how the adopting state is incorporating the innovations into its organizations, practices, and policies. This involves describing any organizational and doctrinal changes occurring, and any variation among the different services in progress on the RMA. Mapping the results of diffusion and innovation within states and organizations reveals indigenous patterns and the range of possible adaptations.[8]

Innovation occurs in three phases—speculation, experimentation, and implementation—and can be gauged by a set of indicators proposed by Mahnken.[9] Speculation involves publishing concept papers, books, journal articles, speeches, and studies regarding new combat methods; forming groups to study the lessons of recent wars; and establishing intelligence collection requirements focused upon foreign innovation activities. Indicators of experimentation include the existence of an organization charged with innovation and experimentation; establishment of experimental organizations and testing grounds; field training exercises to explore new warfare concepts; war gaming by war colleges, the defense industry, and think tanks regarding new warfare areas; and experimentation with new combat methods in wartime. Implementation requires a formal transformation strategy; new units to exploit and counter innovative mission areas; revising doctrine to include new missions; establishing new branches and career paths within the military; changing the curriculum of professional military education institutions; and field-training exercises to practice and refine concepts.

This study addresses innovation and diffusion because these are closely linked, particularly in a globalized world where technology, information, and people move easily across national borders and where the media broadcasts globally and in near-real-time military events that can produce powerful demonstration effects. Diffusion of new technologies, ideas, and practices from abroad is often a key stimulus for innovation. The country studies document how U.S. writings on the RMA have infiltrated discussions of innovation throughout Asia. Diffusion and innovation, however, are not identical. Diffusion focuses on the international or strategic interaction dimension of military change. Innovation focuses on the domestic and organizational dynamics of military change.

MOTIVES AND MODELS

The first step in understanding state responses to the emergence of an RMA is to identify the motives for diffusion. The extant literature posits four types of explanations: security; political economy; technology; and institutional.[10]

Competition is usually assumed to be the major driver of military diffusion. The competitive logic governing the international system creates a powerful incentive for actors to adopt the military practices of the most successful states in the system.[11] "More than any other institution, militaries tend to copy one another across state borders, and with good reason. War is a matter of Darwinian dominance or survival for states, and of life or death for individuals. When an army confronts new or different weaponry or practices on the battlefield, it must adapt to them, and often adaptation takes the form of imitation."[12] While states have a powerful incentive to adopt innovative military methods, full emulation may not be the most efficient way to provide security given their geography, particular factor endowments, demographic pressures, and strategic circumstances.[13] Given mission requirements, however, efficiency is assumed to drive model selection.

Two other security-related explanations are spheres of influence and alliance obligations. If a nation is in a sphere of influence, it may emulate the practices of the bloc leader as a political statement of solidarity. If a nation is a member of a military alliance, it may emulate to facilitate interoperability or specialize to facilitate complementarity.

Political economic explanations focus on the economic pressures to adopt new military practices, emanating from the military–industrial complex, national defense community, or commercial sector. Today, diffusion occurs via commercial as well as military and political channels, and is no longer a state-to-state process managed by central decision-makers to secure the national interest. Firms, organizations, educational institutions, and individuals all play important roles in the transmission of new knowledge and applications.

Technological explanations focus on characteristics of the innovation that encourage or discourage its adoption. Most of the key technologies underlying the current RMA are driven by the civilian commercial economy. There is tremendous commercial pressure for them to spread because they provide a competitive advantage in the global economy.

Two types of institutional explanations are relevant to this study. Bureaucratic interpretations of military behavior focus on inter- and intra-organizational competition and infighting. Although military organizations have a rational interest in adopting the most effective methods to secure the state, they are as likely to be driven by the goal of bureaucratic survival. If an innovation poses a major threat to the organization's missions, resources, autonomy, or essence, it may be strongly resisted; innovations that pose no such threats are far more likely to be adopted. Bureaucratic theorists predict incremental change due to organizational inertia.

Neo-institutional approaches focus on noncompetitive pressures that motivate members of a profession to emulate one another across borders. Through educational and professional networks, organizations share ideas about the best organizational structures and the most legitimate way to practice their profession. Forms or practices sanctioned abroad increase their likelihood of becoming a model for emulation, and international norms exert a powerful influence on national military organizations.[14] Paradoxically, no compelling strategic necessity may be required for diffusion. Neo-institutionalist assumptions predict U.S. models should be preeminent targets of emulation today, just as Prussia was considered a "paradigm" military in the past.[15] There may also be alternative models within a state's region or cultural affinity group, just as the Japanese or "Asian" model of economic development based on state-propelled export-led growth has been far more attractive to regional actors than the "Anglo-Saxon" model.[16]

Motives define the diffusion trajectory and innovation path pursued by the state. However, a variety of forces can intervene to affect the ability of the state to reach the trajectory's endpoint. In the end, the diffusion driver will be one influence, albeit a very important one, on the state's innovation path. Correctly assessing this influence is the first step in the methodology of diffusion diagnostics. It tells us the endpoint to which state leaders aspire. Diffusion drivers and influences are summarized in Table 1.1.

Table 1.1 Drivers of and influences on diffusion and innovation

Driver	Military influenced by
Competition	Most successful models given their geography and factor endowments
Spheres of influence	Bloc leader's models
Alliance obligations	Alliance leader's models
Economic pressures	Models that build on national industrial and commercial strengths
Technology/ commercialization	Models that confer competitive advantage in the civilian commercial economy
Bureaucratic survival	Models that support existing organizational preferences and/or offensive models
Socialization	Most accessible and familiar models
Legitimacy	Most legitimate models as defined by domestic elites and societies

ENABLERS AND CONSTRAINTS

Once motives and models have been identified, the next step in understanding responses to the emergence of an RMA is to assess whether the adopting state can absorb and implement new technologies and practices. Different factors will operate across cases but it is possible to generalize about each factor's impact.

Factors that enable and constrain innovation and diffusion derive from qualities of the innovation and qualities of the adopter. Hardware is often easy to acquire, while software (e.g., doctrine, tactics, and organizational form) is more difficult to develop and implement. Today leading-edge technologies are cheaper and more available than in the past and do not pose the same barriers that precluded states from modernizing their militaries in the industrial age. Nonetheless, developing the software to leverage the hardware may be the more crucial factor in assessing the true capacity of a state's military. The British, Soviet, and American armies all possessed tanks, yet each had difficulty developing the doctrine and organization to wage combined-arms armored warfare.[17] Importing or developing the tactics, doctrine, and training apparatus is *usually* the harder step although in some cases, defense procurement's linkages to domestic economic concerns can complicate the diffusion of hardware. Equipment standardization in NATO is a highly sensitive issue because it requires the purchase of foreign systems or domestic manufacture under license, both of which threaten indigenous defense industries.[18]

The attributes of adopting states that affect diffusion and innovation fall into four categories: the state's polity, economy, society and culture, and military.

Polity

The political environment includes an array of factors from state structure and power, to elite buy-in and commitment to reform, to the relationship between civilian and military authorities.

State structure influences state power and the state's ability to respond to international pressures.[19] State power, or strength, affects what portion of national power governments can extract. It includes capacity (e.g., the ability of the state to extract wealth); coherence (e.g., the extent of centralization or competition among key agencies and branches of government); scope (e.g., how broadly the state defines its responsibilities); and autonomy (e.g., to what extent the state, rather than societal interests, articulate national goals).[20] Hoyt surmises that states with strong state structures will compete successfully in RMAs because they will be able to acquire the necessary financial and human resources.[21]

State structure also affects access to policy-makers and the ease of building winning coalitions for reform. Although a strong centralized state facilitates implementation once a policy decision has been made, centralized state-dominated structures provide reformers with few access points into the political system. Decentralized society-dominated domestic structures provide greater access for reformers but make coalition building difficult because more veto players can derail reform.[22] Most theorists concur that centralized systems inhibit innovation,[23] even though centralization is important for reform if decision-makers are committed to change. Foster and Goodman argue that the "Chinese government is by far the greatest enabler of change" in promoting internet diffusion "through its investments in the educational community...and its management of China Telecom."[24] The ability of reformers to build coalitions and co-opt potential opponents highlights the role of norm entrepreneurs, their access to the decision-making apparatus, and ability to redirect investments into new warfare areas.[25]

Networks developed between domestic and transnational or international actors play a vital role in transferring new ideas and ensuring they are internalized in domestic practices[26] while domestic advocacy groups assist in mobilizing support for new ideas at home.[27] Domestic distributional issues, however, can affect which practices are adopted. Avant argues that "the construction of a coalition around new ideas is most likely to occur when divergent interests or ideas are represented in the dominant coalition."[28] When the dominant coalition shares common ideas and interests and sees new ideas as a threat, old solutions will prevail. When dissent is present in the dominant coalition, reform is more likely. In his analysis of the railroad, rifle, and telegraph revolution, Showalter similarly argues that innovation is stimulated by controversy more than consensus in the military community.[29]

A state's legal and regulatory framework also affects innovation. Regimes that protect intellectual property rights enable innovation. Political norms often rooted in culture and history may inhibit innovation. The major example is Japan, whose constitution renounces war as an instrument of policy. This imposes severe constraints on the country's security policy and is reinforced by legal rulings and tacit political understandings that constrain the growth and activities of the military.[30] Many Japanese equate Article 9 with democracy itself. To date, revision of the constitution has been political taboo.

The political role of the armed forces also influences innovation. External pressures and strategic threats promote competition and spur states to adopt cutting-edge military methods. Internal security roles have the opposite effect.[31] In his assessment of the RMA in the Middle East, Eisenstadt argues that "one of the most important impediments to achieving an RMA...will be political. Nearly every military in the region has an internal security role, and each army has praetorian units organized primarily to counterbalance the regular military and prop up the regime....[T]hese units get the best in equipment and training."[32]

Biddle and Zirkle focus on political instability and internal violence that prompts civilian intervention into military affairs, and reduces the military's ability to effectively use advanced technology.[33] The threat of military violence

Table 1.2 Political factors affecting diffusion and innovation

Factor	Enabler	Constraint
State structure	Centralized, strong	Diffuse, weak
Political diversity	Diverse interests in dominant coalition; controversy in military organization	Consensus in dominant coalition; consensus in military organization
Legal and regulatory framework	Protection of intellectual property rights	Constitutional and legal prohibitions on military activity
Security focus of armed forces	External security focus drives quest for competitiveness and superiority	Internal security focus diverts best troops and resources to prop up regime
Civil–military relations	Professional autonomous military	Politicized military with civilian intervention

within regimes produces pathological civil–military relations where civilian attempts to control the military undermine morale, and incentives and opportunities to learn. Repression inhibits officers' exposure to foreign training; undermines performance because promotion is politically oriented rather than performance-driven; undermines integration and rapid responsiveness because command lines are multiple; and inhibits the accumulation of knowledge and know-how due to rapid rotation, promotion, frequent purges, and suppression of horizontal communications within the military hierarchy. "Where the threat of political violence is low, civil authorities can afford to relax such draconian control measures and in the process make possible a much more effective use of technology by the military."[34]

Table 1.2 summarizes the political factors hypothesized to affect diffusion and innovation.

Economy

Three types of economic factors influence diffusion and innovation: economic growth; industrial and technological capabilities; and defense spending. Economic growth is often linked to a state's long-term potential to enhance its military power. Defense spending may reflect commitment to reform, but it impacts the pace more than the substance of reform. A large defense budget can cover near- and long-term goals, hastening the pace at which reform can occur, but the shape of defense investments is as important for innovation as their magnitude. France enjoyed a clear lead over Germany in military expenditures for nearly the entire interwar period, yet Germany transformed its military to execute *blitzkrieg*.

The state's defense technological and industrial base includes its current capabilities and its ability to improve those capabilities through internal or external means. Internal means require the development of indigenous defense industries, based on local talent, expertise, and research and development (R&D). Taiwan stresses "self-reliant defense" in order to reduce its dependence on foreign suppliers[35] while China strives for a high degree of self-sufficiency in defense acquisition. Indigenous development benefits from protection of intellectual property rights.

External sources of advanced technology include direct transfers of military technology from abroad, purchases of advanced components and equipment from world commercial markets, and technology diffusion from the state's civilian industries that have access to foreign technology and investment. Buying off the shelf allows a state to obtain sophisticated equipment quickly, but the risk is that arms sales can be suspended. China has achieved a high degree of self-reliance in arms production, "one of the developing world's few producers of a full range of military systems"[36] but its current military technology is based on 1950s-era of Soviet technology. China will have to turn to external sources of technology. Limitations imposed by foreign governments on military transfers to China, coupled with the openness of China's civilian industry to foreign technology and investment, makes China's civilian industry the most promising source of knowledge and capability for its defense industry.[37] Accordingly, the level of commercial technological capacity could be an important indicator of military technological capacity provided there is sectoral interconnectedness. The ability to exploit emerging technologies requires both integration into the global economy and a vibrant indigenous technological capacity.

The size of a nation's information industry has been linked to its development into an information society that in turn is linked to its military's transition from an industrial to an informational one.[38] Indicators used to assess the size of a nation's information industry include business volume of telecommunication services; capacity of public switchboards; telephone services; length of optical cable; digital microwave and satellite communication systems; percentage of population that possesses personal computers; and flows of information based on extensiveness of information networks and internet diffusion.[39] Taiwan, for example, has a growing commercial high-tech sector in electronics and information technology (IT), and a highly educated workforce, all of which, analysts argue, indirectly aids its defense industrial base.[40]

Economic arguments about technological capacity often do not distinguish hardware, or the *techne* involved, from software, or the organizational or human application component of a technology. New inventions can be put to use in various ways and often lead to changes in human behavior as their advantages become clear through use. This vital distinction points to the fundamental issue of the organizational, cultural, and societal basis for the introduction, application, and institutionalization of new technologies and practices. In their analysis of China's capacity to adapt and exploit the current RMA, Gill and Henley discuss the underlying organizational structure, incentives, and methods of China's industrial production.[41] China suffers from low interconnectedness, high formalism, and low organizational slack. The military and commercial sectors are segregated, which inhibits

Table 1.3 Economic factors affecting diffusion and innovation

Factor	Enabler	Constraint
Economic growth	Strong	Weak
Defense spending	High	Low
Industrial and technology base	Integration with global economy; indigenous R&D; strong information industry	Norm of self-reliance; dependence upon imports and reverse engineering; weak information industry
Sectoral interconnectedness	Horizontal integration of defense and commercial sectors; free flow of information	Segregated defense sector; high secrecy
Production incentives	Spin-on focus	Commercial focus
Organizational slack	Market economy	Planned economy
Technology transfers	Low export controls on receiving state	High export controls on receiving state

cross-fertilization and diffusion of commercial technologies and organizational principles to the defense sector, and the ability of the military to benefit from spin-on of locally available commercial technology. Bureaucratic formalism pervades organizational norms so meeting production quotas is valued over innovation. Central planning reduces organizational slack and surplus capacity for producers to innovate outside the "plan." Chinese defense industries also rely on "copy production" or reverse engineering, which is increasingly difficult with sophisticated digital technologies. Finally, the incentives in the production of dual-use technologies are for lucrative commercial applications and markets, not spin-on efforts to support military modernization.[42]

Table 1.3 summarizes the economic factors hypothesized to affect diffusion and innovation.

Society and Culture

RMA scholars emphasize that innovation depends as much upon restructuring concepts and organizations as on developing or gaining access to the requisite technologies. Social and cultural factors are critical to these processes. New technologies do not exist in a cultural vacuum. They are not neutral instruments utilized uniformly anywhere, anytime, by anyone. Many case studies of the diffusion of past military innovations demonstrate that innovations requiring significant changes in socio-cultural values and behavioral patterns spread more slowly, less uniformly, and with more unpredictable outcomes.[43]

Rosen's analysis of the armies of India shows how dominant social structures affect the military's ability to generate power.[44] High levels of internal social conflict and fragmented societies, mirrored within militaries, reduce the military power that can be generated from a given amount of resources. His analysis suggests that highly divisive societies will have difficulty in generating military power even if they do have access to advanced technologies. High social conflict may reduce the ability of militaries to innovate and absorb new ideas and technologies for one of several reasons: the military reflects divisions in society and is therefore politicized as Rosen argues; or social conflict thrusts the military into a political role, which then produces pathological civil–military relations that undermine the military's ability to exploit advanced technology as Biddle and Zirkle argue; or internal conflict requires the military to assume a domestic policing role that prevents it from modernizing for external war, as Eisenstadt argues.

Human capital theory focuses on social characteristics such as literacy rates, the education level of the population, and familiarity with machines and electronics, all of which can affect a population's capacity to effectively master and utilize advanced technologies. Human capital theory has been used to argue that "Third World states . . . are at a systemic disadvantage relative to the developed countries in using sophisticated weapons effectively."[45] Demchak posits the importance of wider societal receptivity to networked computers for absorbing the current RMA.[46] Implementing the current RMA "requires appropriate social infrastructure. . . . Wider societal familiarity with networked computers is key to the organization's receptivity in terms of members' knowledge bases which reduces training costs, enhances the likelihood of innovation and lowers the scarcity wage that has to be paid to compete with the wider society for such skills."[47]

Human capital arguments are not uncontested. Cliff, in his discussion of China's human capital, concludes that "absolute numbers of scientists and engineers may be more important than numbers as a proportion of total population, and in this regard China compares more favorably with other countries."[48] Arnett concurs that the human capital of the entire population is less relevant than whether the society can sustain a high-tech sector and whether scientists and engineers are effectively recruited from it into the military. Westney assesses social capacity in terms of the "organizational set" that supports the innovation. Assimilation of new technologies and practices may be problematic if the necessary supporting organizations, such as schools or industries, are inadequately developed.[49]

Culture is typically defined as "worldviews and principled ideas—values and norms—that are stable over long periods of time and are taken for granted by the vast majority of the population."[50] Cultural factors are widely cited as critical to diffusion. Young argues that diffusion is facilitated when nations share common values and language, presumably because this facilitates the transmission of ideas.[51] Hall and Ikenberry note that common cultural heritage of the European states may explain the ease of diffusion of policy innovations among them.[52] Eisenstadt and Pollack conclude from their study of Arab militaries that a society's culture helps determine which skills and behavioral predilections the nation's manpower will bring to military service.[53] Arnett believes that cultural constraints may be the "dominant inhibiting factors" that affect the "design, production and maintenance of weapon systems as much as they do operations."[54] Cross-cultural transfer is rarely complete whether due to imperfect information, the influence of alternative implicit models based on past experience, conflict between the imported model and local patterns, or a different societal scale (e.g., population or geographic area) between the receiving society and the society in which the model originally developed.

Local cultural models often pose barriers to diffusion.[55] Checkel argues the impact of new ideas will be greater if they resonate with domestic norms, understandings and beliefs, if, in his words, there is a "cultural match."[56] Diffusion should be more rapid when a cultural match exists.[57] It makes intuitive sense that new ideas must be compatible with worldviews embedded in political culture or held by elites powerful enough to build winning coalitions. But most culture arguments have difficulty accounting for change if culture by definition is stable.

Goldstone offers an alternative cultural theory of diffusion that views culture as a diverse amalgam of values and beliefs developed, and accreted over time. "Any society may be thought of as having not a single culture, but a variety of cultural 'themes' .... At any one time, political and cultural leaders may adopt a laissez-faire attitude to such cultural diversity .... Cultural diversity and ferment seems likely to favor innovation, and tolerance of pluralism to enhance openness to taking risks, while enforcement of a state orthodoxy seems likely to result in repetition and elaboration of old models, and hostility to innovation and risk."[58] When elites view their role as defenders of a cultural orthodoxy, ideas and practices from abroad, which challenge that orthodoxy will be resisted. Adaptation will be low and only reforms that restore that tradition will be implemented. This thesis resonates with arguments made by Risse and Ropp that socialization [or the internalization of external norms] works well in "open" domestic societies, meaning "societies that for a variety of historical reasons have developed cultures and institutions that are responsive to and can accommodate some meaningful degree of internal debate and external influence."[59]

The business literature provides one more cut on the link between culture and innovation. Based on survey research in 60 countries, Hofstede measured national cultural characteristics according to four dimensions: (1) power distance (short or long), or the level of inequality between people in terms of their power relations, especially the hierarchical boss–subordinate relationship; (2) uncertainty avoidance (high or low); (3) level of individualism (high or low); and (4) masculinity (high or low).[60] A culture characterized by short-power distance tends to produce flatter organizational pyramids; a culture characterized by high uncertainty avoidance tends to have more written rules and micromanagers.

Hofstede's dimensions could be used to assess the impact of culture on innovation. If the IT-RMA military requires a greater degree of delegation of discretion from higher to lower levels of command, a culture characterized by short-power distance may be more desirable. Countries in which political elites try to control the spread of IT for fear they will diffuse power away from central authorities have "control" cultures or a long-power distance. Eisenstadt argues that "regimes that have fought the dissemination of IT in the civilian sphere (note the banning of satellite dishes in Syria, Saudi Arabia, and Iran) are likely to regard with caution military IT that may have the effect of diffusing military power by providing senior commanders a clearer picture of the status and disposition of not only the enemy's armed forces, but their own armed forces as well. They are likely to carefully control the dissemination of these technologies and use them in very different ways than the United States will, in order to—first and foremost—reinforce the regime's control of the military."[61]

A high level of individualism may undercut the degree of integration between different military units (synergy or jointness). A low level of masculinity may ensure greater participation of qualified women in order to enlarge the potential pool of recruitment for an IT-RMA force whose tasks are becoming increasingly more intellectually demanding. Finally, a high level of uncertainty-avoidance, which encourages micromanagement, may not be desirable, given that speed (prompt response) in a rapidly changing war situation is crucial for future warfare.

Table 1.4 summarizes the social and cultural factors hypothesized to affect diffusion and innovation.

Table 1.4 Social and cultural factors affecting diffusion and innovation

Factor	Enabler	Constraint
Social structure	Unified social structure or unifying ideology	High levels of internal social conflict
Human capital	High level of technical education and literacy; societal familiarity with, and use of, computers	Low level of technical education and literacy; low societal familiarity with, and use of, computers
Organizational set	Strong	Weak
Cultural resonance	Strong resonance eases transmission and enhances desire for adoption	Weak resonance inhibits transmission and diminishes desire for adoption
Cultural tolerance	Tolerance of diversity and internal debate facilitates innovation and diffusion	Official orthodoxy hinders innovation and diffusion
National culture	Participatory	Control
	Short power distance	Long power distance
	Low uncertainty avoidance	High uncertainty avoidance
	Low individualism	High individualism
	Low masculinity	High masculinity

Military Organizations

Military organizations can be seen as natural systems, rational systems, and open systems. As a natural system, military organizations strive to survive and protect their self-interests in an environment of scarce resources and internecine strife. This produces a tendency to be conservative and risk averse, and to adopt only technologies and strategies that will defend and enhance the organization's resources, autonomy (jurisdiction and independence), and organizational essence (the views on missions and capabilities held by the dominant group in the organization).[62] Change is incremental and adjustments consistent with existing tasks. Organizations place a premium on predictability, stability, and certainty. These values are inimical to innovation, although "innovations that pose no threat to organizational routines, strategies or essence are often readily adopted. It is only the new weapons that portend major organizational changes, reallocation of resources, the possibility of diminished organizational autonomy and so forth that meet resistance."[63]

It is logical to assume from the natural systems perspective that innovation, which necessarily entails a major restructuring of the military organization, will not be easy. Not only is inertia inherent in the functioning of any large organization; dominant group interests can network in the policy environment to support their interests. As Rosen argues, organizations are complex political communities and innovation is an ideological struggle over a new theory of victory. Implementation of an innovation requires "creating new career paths along which younger officers specializing in the new tasks could be promoted."[64] Rosen's analysis suggests that change is possible if the distribution of power within the organization is not highly skewed toward a particular branch with strong legacy system interests. The issue may be less that of organizational conservatism per se and more a distributional question of whether those groups that are conservative are also dominant in the organization's power structure. If the distribution of power in the organization is more balanced, then pockets of transformers are as likely to enable change, as traditionalists are to hinder change.

As a rational system, military organizations strive to improve the efficiency with which they secure the state.[65] They respond rationally to the dictates of strategic geography, technological developments, and enemy behavior.[66] They reevaluate beliefs, redefine tasks, and learn from experience.[67] The conditions that promote learning include pressure from civilians or other military actors, which provide "the impetus, political incentive, and political opportunity for a significant reevaluation of assumptions";[68] the existence of credible knowledge and experience that supports the innovation; and urgency or pressures from the international environment. Urgency aids problem identification and increases pressure on the organization to focus its strategic priorities and bring experience to bear on its strategic problem. Krepinevich concludes that certain military organizations are better able to exploit the advantages of military revolutions than are others because they are able to focus more precisely on specific contingencies and competitors.[69]

Table 1.5   Military and organizational factors affecting IT-RMA diffusion and innovation

Factor	Enabler	Constraint
Existing organizational preferences	Parity in power among service's branches	Asymmetry in power among service's branches skewed toward legacy systems
Domestic pressure	High and multiple sources	Low
Experiential base	Strong	Weak
International vulnerability	High	Low
Organizational type	Cybernetic; rational, learning system	Socio-political; military highly politicized
Organization's beliefs	Meshes with innovation	Conflicts with innovation (e.g., ANZAC spirit)
Interconnectedness	High promotes jointness	Low feeds inter-service rivalry

Demchak also argues that learning or "cybernetic" organizations are the most supportive for military innovation. They are proactive and change in response to "sensors consciously designed to monitor stimuli from the external environment."[70] Cybernetic organizations can be much more innovative than "organic" organizations, which evolve to enhance their own survival, or "social" organizations, which are constrained by the deeply held beliefs of their members. Organizations will accept risk and can overcome inertia in order to achieve the most desirable result in the outside world.

As an open system, military organizations are manifestations of powerful institutional rules (or beliefs, understandings, and standards about the ways things ought to be) and myths that are binding on their members.[71] Militaries can change but change is shaped by beliefs collectively held by members of the organization. Those beliefs may be rooted in recent experience or deep historical practice. To assess an organization's potential for innovation, it is necessary to look at the way the organization identifies itself, the enemy, the nature of warfare, and the appropriate way to wage war. Organizations also constitute their environments in that particular technologies succeed because of the social networks that support them. Military change does not simply follow in the wake of new technologies unless the technology is promoted by scientists and powerful social networks.

Table 1.5 summarizes the military and organizational factors hypothesized to affect diffusion and innovation.

SPREAD OF THE INFORMATION RMA

The extent to which the United States will be able to enjoy its military lead depends on whether and how others assimilate and exploit the innovations associated with the IT-RMA. Understanding diffusion and the dynamics of innovation abroad is essential to U.S. policy choice and strategy.

The spread of revolutionary military innovations across the international system raises a series of questions that must be answered if we are to understand how the IT-RMA is likely to spread and transform militaries worldwide. First, when do states attempt to adopt RMA innovations and transform their militaries? Always, only when severely threatened or when exposed to a highly salient demonstration, or not always even then? Second, are most attempts at innovation carried through to completion, or are they often blocked from full implementation? Third, when innovation occurs, is it faithful to any particular imported model or is the result a hybrid quite different from either the source's or the state's own prior practice? Fourth, what are the barriers to innovation that explain the answers to these questions, and to what extent can states overcome these barriers?

The capacity to assimilate new technologies, doctrines, and behavioral practices is one of the most important dimensions of the diffusion equation that will affect the scope and pace of innovation and diffusion. Capacity has political, economic, social, cultural, and organizational dimensions. Each plays

Figure 1.1 Key enablers of diffusion and innovation.

a role in fostering innovation in the public and private sectors, and in facilitating the flow of ideas across borders, sectors, organizations, and services. Figure 1.1 distills from the preceding discussion a set of key factors that enable diffusion and innovation. Scoring any particular country across these factors will provide an overall assessment of their ability to exploit the RMA.

NOTES

John Kennedy, Satoshi Shimada and Leo Blanken provided valuable research assistance for this project.

1. See Emily O. Goldman and Andrew L. Ross, "The Diffusion of Military Knowledge: Theory and Practice," in Emily O. Goldman and Leslie C. Eliason, eds., *The Diffusion of Military Technology and Ideas* (Stanford, CA: Stanford University Press, 2003), pp. 371–403.
2. Paul Dibb, "The Revolution in Military Affairs and Asian Security," *Survival*, 39:4 (Winter 1997–98).
3. David Roessner and Michael Salamone, "National Technological Competitiveness and the Revolution in Military Affairs," final report, phase II, prepared for the Director, Office of Net Assessment, Office of the Secretary of Defense, June 1, 1999. Notably, Roessner and Salamone did not include Australia and Japan in their study.
4. Stephen Biddle, *Afghanistan and the Future of Warfare: Implications for Army and Defense Policy* (U.S. Army War College, November 2002).
5. My thinking on this benefited from conversations with Steve Biddle.
6. U.S. Department of State, *Patterns of Global Terrorism 2002*, Appendix H.
7. Bruce Berkowitz, "Keeping the Lid on Pyongyang," *Wired*, 11:6 (June 2003).
8. This methodology is proposed in Goldman and Eliason, eds., op. cit.
9. Thomas G. Mahnken, "Uncovering Foreign Military Innovation," *Journal of Strategic Studies*, 22:4 (December 1999).
10. Chris C. Demchak, "Creating the Enemy: Global Diffusion of the Information Technology-Based Military Model," in Goldman and Eliason, eds. (2003) op. cit., p. 316.
11. Kenneth N. Waltz, *Theory of International Politics* (Reading, MA: Addison-Wesley, 1979).
12. John A. Lynn, "The Evolution of Army Style in the Modern West, 800–2000," *The International History Review*, 18:3 (August 1996), p. 509.
13. Colin Elman, *The Logic of Emulation: The Diffusion of Military Practices in the International System* (Ph.D. Dissertation, Columbia University, 1999).
14. See Theo Farrell, "World Culture and the Irish Army, 1922–1942" in Theo Farrell and Terry Terriff, eds., *The Sources of Military Change: Culture, Politics and Technology* (Boulder, CO: Lynne Rienner, 2002) pp. 69–90; Demchak, (2003), op. cit.; Theo Farrell, "Transnational Norms and Military Development: Constructing Ireland's Professional Army," *European Journal of International Relations*, 7 (March 2001); Dana P. Eyre and Mark C. Suchman, "Status, Norms, and the Proliferation of Conventional Weapons: An Institutional Theory Approach," in Peter J. Katzenstein, ed., *The Culture of National Security: Norms and Identity in World Politics* (New York: Columbia University Press, 1996) pp. 79–113.
15. Lynn, op. cit.

16. Mahathir has an explicit "look east" orientation in his development strategy and uses that very phrase. Korea is a clone of Japan. Malaysia, Korea, Taiwan, and Singapore all have had strong government intervention in economic development.

17. Thomas G. Mahnken, "Beyond *Blitzkrieg*: Allied Responses to Combined-Arms Armored Warfare during World War II," in Goldman and Eliason, eds., op. cit., pp. 243–266.

18. Thomas-Durrell Young, "Cooperative Diffusion through Cultural Similarity: The Post-War Anglo-Saxon Countries' Experience," in Goldman and Eliason, eds., op. cit., pp. 93–113.

19. Stephen Krasner, *Defending the National Interest* (Princeton: Princeton University Press, 1979); Fareed Zakaria, *From Wealth to Power: The Unusual Origins of America's World Role* (Princeton: Princeton University Press, 1998), pp. 35–41.

20. Zakaria (ibid.), pp. 38–39.

21. Timothy D. Hoyt, "The Revolution in Military Affairs and the Developing World: What Can We Expect and Where?" paper submitted for the Annual Convention of the International Studies Association (April 17, 1996) p. 9.

22. Thomas Risse-Kappen, "Ideas Do Not Float Freely: Transnational Coalitions, Domestic Structures, and the End of the Cold War," *International Organization*, 48:2 (Spring 1994), pp. 185–214; Matthew Evangelista makes a similar argument in his book, *Innovation and the Arms Race: How the United States and the Soviet Union Develop New Military Technologies* (Ithaca: Cornell University Press, 1988).

23. Evangelista, op. cit., pp. 29–33.

24. William Foster and Seymour E. Goodman, *The Diffusion of the Internet in China* (September 12, 2000) p. 78 (http://www.public.asu.edu/~wfoste1/chinainternet.pdf).

25. Farrell (2001) op. cit., p. 81.

26. Thomas Risse, Stephen C. Ropp, and Kathryn Sikkink, *The Power of Human Rights: International Norms and Domestic Change* (Cambridge: Cambridge University Press, 1999).

27. Margret Keck and Kathryn Sikkink, *Activists Beyond Borders: Transnational Advocacy Networks in International Politics* (Ithaca: Cornell University Press, 1998); Thomas Risse and Kathryn Sikkink, "The Socialization of International Human Rights Norms into Domestic Practices: Introduction," in Risse, Ropp, and Sikkink, eds., op. cit., pp. 1–38; Amy Gurowitz, "Mobilizing International Norms: Domestic Actors, Immigrants, and the Japanese State," *World Politics*, 51:3 (1999), pp. 413–445.

28. Deborah Avant, "From Mercenary to Citizen Armies: Explaining Change in the Practice of War," *International Organization*, 54:1 (Winter 2000), p. 49.

29. Dennis Showalter, "Information Capabilities and Military Revolutions the 19th Century Experience," paper prepared for the CSBA Workshop on "Military Revolutions: The Role of Information Capabilities," March 4–5, 2002, Washington, D.C.

30. Arthur J. Alexander, "Japan's Potential Role in a Military–Technical Revolution," report prepared for the Office of the Secretary of Defense (Net Assessment), January 13, 1995; Peter J. Katzenstein and Nobuo Okawara, "Japan's National Security," *International Security* (Spring 1993); Katzenstein (1996) op. cit.; Thomas U. Berger, *America's Reluctant Allies: The Genesis of the Political–Military Cultures of Japan and West Germany* (Ph.D. Dissertation, Massachusetts Institute of Technology, 1992).

31. Hoyt, op. cit., p. 9.
32. Michael J. Eisenstadt, "The Future Middle Eastern Threat Environment and the Revolution in Military Affairs (RMA)," manuscript, p. 3.
33. Stephen Biddle and Robert Zirkle, "Technology, Civil–Military Relations and Warfare in Southern Asia," in Arnett, ed., op. cit., pp. 317–345.
34. Ibid., p. 320.
35. Richard A. Bitzinger and Bates Gill, *Gearing Up For High-Tech Warfare? Chinese and Taiwanese Defense Modernization and Implications For Military Confrontation Across the Taiwan Strait, 1995–2005* (Center for Strategic and Budgetary Assessments, February 1996), pp. 33–34.
36. Ibid., p. 16.
37. Roger Cliff, *The Military Potential of China's Commercial Technology* (RAND, 2001), p. ix.
38. Major General Wang Baocun, "China and the Revolution in Military Affairs (1)," *China Military Science*, 4 (2001), p. 148.
39. Foster and Goodman, op. cit.
40. Bitzinger and Gill, op. cit., p. 36.
41. Bates Gill and Lonnie Henley, *China and the Revolution in Military Affairs* (Strategic Studies Institute, 1996), pp. 7–9.
42. Bitzinger and Gill, op. cit., p. 20.
43. Goldman and Eliason, eds., op. cit.
44. Stephen Peter Rosen, *Societies and Military Power: India and its Armies* (Ithaca: Cornell University Press, 1996).
45. Biddle and Zirkle, op. cit., pp. 317–318.
46. Chris C. Demchak, "RMA in Developing States: Botswana, Chile and Thailand," manuscript prepared for National Security Studies Quarterly Conference "Buck Rogers or Rock Throwers? Technology Diffusion, International Military Modernization, and the International Response to the Revolution in Military Affairs," October 14, 1999.
47. Ibid., p. 21.
48. Cliff, op. cit., xii.
49. D. Eleanor Westney, *Imitation and Innovation: The Transfer of Western Organizational Patterns to Meiji Japan* (Cambridge: Harvard University Press, 1987), pp. 28–31.
50. Risse-Kappen, op. cit., p. 209.
51. Thomas-Durrell Young, "Cooperative Diffusion Through Cultural Similarity: The Postwar Anglo-Saxon Countries' Experience," in Goldman and Eliason, eds., op. cit., pp. 93–113.
52. John A. Hall and G. John Ikenberry, *The State* (Milton Keynes, UK: Open University Press, 1989).
53. Michael J. Eisenstadt and Kenneth M. Pollack, "Armies of Snow and Armies of Sand: The Impact of Soviet Military Doctrine on Arab Militaries," in Goldman and Eliason, eds., op. cit., pp. 63–92.
54. Arnett, op. cit.
55. Theo Farrell, "Culture and Military Power," *Review of International Studies* 24 (1998); Jeffrey W. Legro, *Cooperation Under Fire: Anglo-German Restraint During World War II* (Cornell University Press, 1995); Peter J. Katzenstein, *Cultural Norms and National Security: Police and Military in Japan* (Ithaca: Cornell University Press, 1996); Elizabeth Kier, *Imagining War: French*

and British Doctrine Between the Wars (Princeton: Princeton University Press, 1997).

56. Jeffrey T. Checkel, "Norms, Institutions and National Identity in Contemporary Europe," *International Studies Quarterly*, 43 (1999), pp. 83–114.

57. Thomas Risse and Stephen C. Ropp, "International Human Rights Norms and Domestic Change," in Risse, Ropp and Sikkink, eds., op. cit., p. 271.

58. Jack A. Goldstone, "Political Crisis and Cultural Orthodoxy: Attitudes toward Risk and Innovation in the Divergence of East and West in the 17th Century," draft chapter, 378.

59. Risse and Ropp, op. cit., pp. 262–264.

60. Geert Hofstede, *Culture's Consequences*, 2nd. Ed. (Beverly Hills: Sage Publications, 1994).

61. Eisenstadt, op. cit., p. 3.

62. Richard Cyert and James March, *A Behavioral Theory of the Firm* (Englewood Cliffs, NJ: Prentice-Hall, 1963); Herbert Kaufman, *The Limits of Organizational Change* (University, AL: University of Alabama Press, 1971); James Q. Wilson, *Bureaucracy: What Government Agencies Do and Why They Do It* (New York: Basic Books, 1989); Edward L. Katzenbach, "The Horse Cavalry in the Twentieth Century: A Study in Policy Response," Public Policy 7 (1958), pp. 120–149.

63. Evangelista, op. cit., pp. 11–12.

64. Stephen Peter Rosen, *Winning the Next War: Innovation and the Modern Military* (Ithaca: Cornell University Press, 1991), p. 76.

65. Samuel P. Huntington, *The Soldier and the State* (Cambridge: Harvard University Press, 1957).

66. Rosen (1988) and (1991), op. cit.; Kimberly Martin Zisk, *Engaging the Enemy: Organization Theory and Soviet Military Innovation, 1955–1991* (Princeton: Princeton University Press, 1993).

67. Jack S. Levy, "Learning and Foreign Policy: Sweeping a Conceptual Minefield," *International Organization*, 48:2 (Spring 1994), pp. 287–288; Barbara Levitt and James G. March, "Organizational Learning," *Annual Review of Sociology*, 14 (1988), p. 324.

68. George Breslauer, "Ideology and Learning in Soviet Third World Policy," *World Politics*, 39 (April 1987), p. 443.

69. See Andrew F. Krepinevich, "Cavalry to Computer: The Pattern of Military Revolutions," *The National Interest* (Fall 1994), p. 39.

70. Demchak (1999), op. cit., p. 13.

71. Theo Farrell, "Figuring Out Fighting Organizations: The New Organizational Analysis in Strategic Studies," *The Journal of Strategic Studies*, 19 : 1 (March 1996), pp. 124–125.

2

AUSTRALIA'S APPROACH TO THE REVOLUTION IN MILITARY AFFAIRS, 1994–2004[1]

Michael Evans

We have got to have an information edge. We have just seen 21st-century warfare.

General Peter Cosgrove, Chief of the Australian
Defense Force on the Iraq War, May 1, 2003

Since the late 1990s, Australian defense planners have confronted the reality that the globalization of security means that national and international approaches to defense can no longer be easily separated. The long period in which Australia could rest its defense policy upon protecting its own strategic geography came to an end, initially with Australian intervention in East Timor in 1999, and then, with the terrorist attacks of September 11, 2001 on the United States. After the tragedy of September 11, Australia invoked the Australia, New Zealand, and the United States (ANZUS) Alliance and between 2001 and 2003 deployed forces first to Afghanistan and then to Iraq. Australia has become a key U.S. ally in an ongoing global war on terror— a conflict that has led Canberra not only to commit combat forces for overseas deployments, but also to focus on new requirements for homeland security, including the creation of a Special Operations Command.

The merging of Australia's national, regional, and global security concerns has been further underlined by events in Bali and the Solomon Islands during 2002 and 2003. For Australians, the Bali terrorist bombing of October 12, 2002 saw the country suffer its greatest single loss of civilian life since the Japanese bombing of Darwin in 1942.[2] The Bali tragedy brought together many of the non-state elements of the new era of globalized security. The attack involved South East Asian operatives from Jemaah Islamiah with close links to the internationalist al-Qa'ida movement. In attacking Western nationals in the world's largest Muslim country, the Bali conspirators acted as symbolic representatives of a global ideology of radical political Islam.

Similarly, Australia's dispatch of military forces to restore governance to the Solomon Islands in July 2003 was based on an acknowledgment that modes of security have merged. Central to Australian policy was an explicit assumption that in an age of globalization, "failed states" act as regional magnets for networks of transnational dissidents and are not in the interests of any democratic country. Intervention in both East Timor and the Solomons represents a proactive policy toward what former Defense Minister, John Moore, once described as a "sea of instability" emerging in the Asia-Pacific.[3] This "sea of instability" embraces the Solomons, Papua New Guinea, and islands such as Bougainvillea, Vanuatu, and Fiji in the South and South West Pacific, insurgency in Ache and Papua in Indonesia, and insurrection in parts of the Philippines.[4]

The threat to Australia's national security posed by a conjunction of regional instability and global terrorism was again emphasized by the present Minister for Defense, Senator Robert Hill, in October 2003. To Moore's notion of a "sea of instability", Hill added the construct of an "arc of terrorism":

> We have sometimes defined Australia's strategic circumstances in terms of an arc of instability...What we have to contend with [now]...is an arc of terrorism stretching from South East Asia into Pakistan and Central Asia, and then spilling out into the Middle East and the Horn of Africa. It [the threat from terrorism] demonstrates the flaws and even the dangers of trying to draw a line around Australia's strategic interests. Regional terrorism, global security and the defense of Australia in this new strategic environment are very much the same thing.[5]

All of the above might suggest that the globalization of security has been overwhelmingly negative for Australian defense policy. Yet globalization is also a positive process for, while it presents complex policy challenges, these challenges have been balanced or offset by its offspring—the Revolution in Military Affairs (RMA) based on information technologies. If globalized security diminishes the relative significance of strategic geography, creates asymmetry, and merges modes of conflict, the RMA provides a range of new technologies whose potential for connecting sensors and networking weapons systems offer advanced states important advantages in developing new military capabilities. Australia is one of these advanced states. With its large landmass, small population, low-birth rate, and, with its prosperity based on regional and global economic networks, Australia stands to gain greatly from the use of information-based high-technology military assets.

Not surprisingly, then, since the mid-1990s Australian defense planners have given considerable attention to the notion that there are practical benefits to be gained from acquiring selected information technologies arising out of the RMA. For many official Australian strategists, RMA developments in information technology represent one of the most important means to redesign Australia's approach to defense planning in the twenty-first century.

Critical issues of military capability, force structure organization, and joint doctrine are seen as having at least partial solutions in the realm of RMA research and development.

This chapter provides an overview of Australia's quest to exploit the RMA in order to strengthen its defense capacity. It concentrates on five areas. First, it describes the background to the rise of RMA thinking in Australia between 1994 and 1997. Second, it examines the manner in which RMA thinking was institutionalized in Australian strategic thought between 1997 and 2000, including the development of an indigenous concept of an information-based military revolution called the Knowledge Edge. Third, this chapter analyzes the significance of the December 2000 Defense White Paper in the Australian process of exploiting selected information-age technologies to achieve information superiority and a Knowledge Edge. Fourth, it assesses some of the major challenges confronting Australia's quest to redesign its armed forces around RMA ideas and technologies, with particular emphasis on the political economy of defense expenditure and the impact of the changing security environment since September 11, 2001. Finally, the study provides a brief analysis of the rise to prominence of network-centric warfare in Australian strategy since 2002 and the implications of this approach in the quest for a Knowledge Edge.

THE BACKGROUND TO AUSTRALIA AND THE RMA: THE ERA OF INFORMAL DEBATE, 1994–97

The Australian Defense Organization (ADO) did not adopt RMA thinking into its official defense policy until the end of 1997. The early years of the Australian RMA debate, the era of informal "first-phase theorizing" have been outlined in considerable detail elsewhere.[6] Nonetheless, it is useful to summarize the main features of the Australian RMA approach in order to understand its character.

Between 1994 and 1997, the Australian RMA debate was largely the work of uniformed officers in the services and of defense scientists who were concerned with analyzing future warfare. Consequently, Australian examination of RMA developments tended to be singular and informal, rather than joint and institutional in approach. Early RMA theorists included Colonel (now Lieutenant General) Peter Leahy, Brigadier (later Major General) Peter Dunn, Air Vice Marshal Peter Nicholson and scientists such as Dr. Richard Brabin-Smith (formerly Chief Defense Scientist and later Deputy Secretary for Strategy).[7] The early theorists concentrated on analyzing the potential benefits of information technology in overcoming the problem of defending Australia—a country covering 12 percent of the earth's surface but containing 1 percent of the earth's population. To put this task into context, it should never be forgotten that Australia's northern frontier extends for the same distance as that between London and Beirut.

The first-phase theorists focused on the roles of command, control, communications, computers, and intelligence (C4I) and command and control warfare (C2W). By 1996, there was a general consensus among senior uniformed professionals such as Leahy, Dunn, and Nicholson that Australian Defense Force (ADF) operations would have to be transformed, as the technological changes of information age warfare became more apparent.[8] A 1996 paper by Air Vice Marshal Peter Nicholson, Air Commander, Australia, saw the key to an Australian RMA response as lying in acquiring sensor suites and data fusion giving improved situational awareness in operations. Nicholson called his approach to the RMA one of "knowledge dominance"—an idea that was subsequently to assume great importance in official Australian defense circles.[9]

The views of the uniformed theorists were supported by the then Chief Defense Scientist, Dr. Brabin-Smith, who argued that Australia stood to benefit in the early twenty-first century from emerging technologies in information, surveillance, and reconnaissance (ISR) communications, C2, and precision strike.[10] A significant practical development in the RMA debate was the decision by the Defense Science and Technology Organization (DSTO) in 1996 to launch the *Takari* Program—a scheme aimed at delivering a viable and integrated C3I capability to the ADF for operations on the battlespace of the future.[11]

Australian thinking on the importance of the RMA was also strongly influenced by exposure to U.S. experimentation. Australian analysts studied programs such as the U.S. Army's Force XXI scheme, its Advanced Warfighting Exercises, digitization program, and its use of battle laboratories. In addition, joint American–Australian military exercises under U.S. Pacific Command demonstrated the use of C4I and battlespace detection systems in improving the speed and efficiency of military decision-cycles.[12]

A decisive event in the development of an official Australian RMA initiative was the election in March 1996 of a Liberal–National Coalition Government led by John Howard. Under Minister for Defense Ian McLachlan, the new administration demonstrated an early interest in the possibilities of RMA technology. In June 1996, McLachlan argued that the long-term changes in information technology would be as profound for military organizations in the twenty-first century as the coming of the internal combustion engine in the early twentieth century.[13] He identified the RMA's key components as being fourfold: lethality of weapons; projecting force over increased distances; speed of information processing; and growing capacities for intelligence gathering.[14] The minister pointed to other benefits such as the potential of unmanned aerial vehicles (UAVs) and increased interoperability with allies. However, he warned that Australia had to be "careful to pick only those parts of RMA technology that address our needs."[15]

By the end of 1996, Australian–American cooperation on the RMA increased dramatically. Australian defense strategists became immersed in the full range of American ideas on information warfare. These ideas included Admiral William Owens' theory of the "emerging systems of systems"; notions of battlespace awareness and dominant maneuver; precision strike, sensor to shooter links and simultaneity; the potential of joint direct attack munitions (JDAM), global positioning systems (GPS), and brilliant sub-munitions.[16] From 1996 onward, American future warfare specialists from the Office of Net Assessment (ONA), the Center for Strategic and Budgetary Assessments (CSBA), and the American war colleges became regular visitors to Australia.

In early 1997, Andrew Marshall, the distinguished American strategic thinker and Director of Net Assessment, pointed out that Australia stood to benefit from several RMA developments. He singled out automated combat systems, long-range precision-strike, and stealth and sensor technology as new techniques that would permit control of Australia's huge northern sea–air gap in a way not possible before. Marshall also thought the U.S. Marine Corps concept of *Sea Dragon*—in which small units operated with logistics and firepower from a distance—might be a useful model for Australia to emulate in terms of projecting power in the future.[17] Significantly, from the beginning of 1997, ONA consultants became influential in helping to mold the Department of Defense's institutional approach to the RMA debate.

AUSTRALIA'S INSTITUTIONAL EMBRACE OF THE RMA, 1997–2000

In December 1997, a new defense review, *Australia's Strategic Policy, 1997* (ASP 97) became the first official document to acknowledge the potential of the RMA in helping Australia to shape its future strategic environment.[18] ASP 97 argued that the application of information technology within the ADF would permit more cost effectiveness in force structure through "exploiting technology, doctrine and geography."[19] The review went on to state:

> For Australia it [the RMA] has particular significance. Not only will new tech-nology provide military personnel with an expansive breadth and depth of information about the battlefield, but sophisticated strike weapons will give advanced forces the capability to destroy targets with an unparalleled degree of precision and effectiveness.[20]

Mastery of information technology would be an area where the small, 50,000 strong ADF could aspire to continuing excellence.[21] Australia's highest capa-bility priority in the future was described as being the achievement of a Knowledge Edge. The Knowledge Edge construct was an apparent refine-ment of Air Vice Marshal Nicholson's earlier concept of "knowledge domi-nance" as well as reflecting the research work of the DSTO.[22] The Knowledge Edge was defined in ASP 97 as "the effective exploitation of information technologies to allow us to use our relatively small force to maximum effectiveness."[23]

Exploiting information age technology to achieve a Knowledge Edge was seen as holding out three important strategic advantages for Australia. First, information capabilities offered the possibility of greatly improved surveillance of Australia's vast maritime approaches. Second, information technology—when applied to the command, positioning, and targeting of forces—would enable military deployment to maximum effect. Information technology offered a means of mastering Australia's geography.[24]

Third, through its strong assets in domestic information technology and its alliance with the United States, the ADF could look forward to creating a defense architecture that integrated the three elements of capability: intelligence, command, and its supporting systems including communications and surveillance.[25] With American assistance, ASP 97 foresaw sensors, platforms, space-based surveillance, long range UAVs, over-the-horizon radar (OTHR) and airborne early warning and control aircraft (AEW&C) being meshed into an overall system to provide comprehensive real-time information to the ADF in the field.[26]

The Office of the RMA and the Futures Directorates: The Establishment of Australia's Future Warfare Organization

During 1998 and 1999, the Howard Government introduced several further measures to support an Australian RMA effort. The DSTO increased spending on RMA-related research and development into C4, ISR, and Electronic Warfare (EW) by A$10 million. In addition, the ADF expanded military cooperation with the U.S. Army's battle laboratories.[27] However, the Government's most important measure was the decision in April 1999 to create the Office of the RMA in the Military Strategy Branch of Australian Defense Headquarters. The formation of a dedicated RMA organization in the heart of Australia's defense machinery insured that what has been called "second phase" theorizing on information age warfare would be more formalized, institutional and above all more triservice in approach.[28]

The Office of the RMA was to be headed by the ADF Director General of Military Strategy, a one star officer, who was to report directly to the Secretary of Defense and the Chief of the Defense Force and through them to the Minister.[29] The main objective of the new Office was to extract "the maximum value from the RMA for the ADF—be it in equipment acquisition and development, training, doctrine development or alliance relations."[30] In particular, the new Office was to seek to identify those aspects of technological change that were most likely to affect major long-term capabilities.

The Office of the RMA became responsible for coordinating three important tasks. First, in close cooperation with the United States, the Office was charged with developing an implementation strategy for adapting selected aspects of RMA technology to Australia's circumstances. Second, the new organization was to identify and analyze future warfare concepts that could be used to incorporate organizational, doctrinal, and technological changes into the current ADF. Third, the Office was to prepare for the Minister for Defense a paper on the ADF and the implications of the RMA exploring policy options and alternatives.[31]

Parallel to the formation of the Office of the RMA, the single services refined their input into the environmental specialties of information age conflict. The army, air force, and navy formed dedicated future warfare directorates to facilitate wider collaboration and cross-pollination in research.[32] In the land environment, the Australian Army's Future Land Warfare Directorate was created in 1999 to examine land warfare trends out to 2030 based on a "concept-led, capability-based" philosophy involving network-centric warfare and battlespace synchronized operations.[33] Similarly, in the RAAF, Project Oracle 2030 was created to try to "pre-adapt the RAAF" to twenty-first-century operations by examining such approaches as effects-based operations.[34] During 2000, the RAN created a Strategy and Futures Directorate to try to fuse together blue-water responsibilities with the growing need in the twenty-first century for integrated operations in the littoral using network-enabled operations and UAVs.[35]

Between 1999 and early 2000, the formation of the Office of the RMA and the creation of the dedicated single service future warfare directorates did much to establish an institutional framework for the disciplined analysis of RMA concepts. The Office of the RMA and the futures directorates also contributed decisively to the notion that there was an affordable way for Australia to absorb and benefit from the rigorous challenges arising from warfare in the information age.

Project Sphinx: Australia's Methodological Approach to the RMA and Future Warfare

Between 1999 and 2000, the Office of the RMA developed a methodological strategy for an Australian approach to information age warfare called Project Sphinx. The project also served to provide Australia with a coordination mechanism to develop concepts for the ADF to meet the needs of warfare in the information age.[36]

Sphinx sought to provide a collaborative methodology to analyze RMA developments. The focus of the project has been on identifying conceptual issues related to capability and doctrinal usage thereby providing a firm intellectual foundation for a research and development effort into RMA-style technology.[37] The overall objective was to use Sphinx to help create what was described as a strategic level Enterprise Architecture Model within the ADO that unites policy, operations, systems, and technical processes. Until the adoption of network-centric warfare in 2002, Sphinx was seen as the means to identify the most plausible future warfare concepts and to assess their possible long-term capability investment implications for Australia through to the year 2025.[38]

Central to Australia's Sphinx methodology were three strategic propositions. First, the Asia-Pacific region is regarded as central to Australia's security. Second, there was a firm Australian belief that the information age has ushered in a new era in warfare. Third, there was a general strategic conviction that the post–Cold War security environment was peculiarly volatile and extremely difficult to predict.[39] Project Sphinx attempted to grapple with the problem of identifying and exploring future warfare concepts and their consequences by employing three processes: *concept generation, concept evaluation*, and *concept consultation*. Concept generation was originally facilitated by the formation of Concept Initiation Teams (CITs). These teams—drawn from wide expertise throughout the Department of Defense—provide a means to assess the impact of emerging information-age warfare techniques.[40]

Throughout 1999 and 2000, CITs examined various categories of future warfare in information age conditions. These categories included ISR, C2, and adaptive interoperability, tailored effects (or precision firepower), force projection, force protection and force sustainment. The aim of each team was to refine concepts that could serve as potential pathways to guide future ADF capability planning and force structure.[41] In 2000, in order to link concept development to capability assessment, a Military Systems Experimentation Branch was created within the DSTO.[42]

The second process in the Sphinx program has been concept evaluation—mainly through the use of campaign wargames known as the Krait strategic seminar series. Strategic wargaming was introduced into the ADO in order to evaluate the feasibility of future warfare concepts in various conflict scenarios that might emerge in the first quarter of the twenty-first century. The Krait process was viewed as important in testing the various warfare concepts in order to establish which ones offered the best possibilities for exploiting military advantage in future joint, combined, and coalition operations planning. The Office of the RMA believed that wargaming would eventually be accepted as an important intellectual exercise in the Australian capability development process.[43]

However, between 1999 and 2000, Australian wargaming relied heavily on American rather than indigenous expertise.[44] The ADO contracted consultants drawn from U.S. organizations such as the CSBA and Science Applications International Corporation (SAIC).[45] In particular, the CSBA's experience in conducting the 20XX Series of futuristic wargames for the U.S. ONA was regarded as particularly valuable by Australian defense planners.

In January 2000, the Military Strategy Branch established a liaison position with U.S. Joint Forces Command for collaboration in future warfare experimentation. The objective of this relationship was to "provide a specialist liaison and representation link between the ADO and US Joint Forces Command on issues related to the RMA." Important emphasis was placed on C4ISR work, operational procedures such as effects-based operations and RMA wargames.[46]

Using largely CSBA methods, Krait wargames modeled several Asia-Pacific conflict scenarios ranging from major war through regional coalition operations to the unilateral use of Australian forces in a "failed-state."[47] In 1999 and 2000, Krait wargames also tested future warfare concepts such as force projection and force protection, ISR, command and control, and force sustainment; tailored effects and special operations. The most recent Krait seminar has involved a workshop on developing a Joint Warfighting Concept for the ADF in information age conditions. The Krait process is supported by another series called Taipan that concentrate on refining campaign concepts and force structure analysis.[48]

The third process in Project Sphinx, the process of concept consultation, was facilitated by the creation in August 1999 of an RMA Working Group. The latter was formed by drawing on the intellectual resources of the Department of Defense, academia, and industry to help refine Australia's future warfare concepts. The initial RMA Working Group included an eclectic collection of policy-makers, defense analysts, research scientists, uniformed professionals, academic consultants, and representatives from private industry. During 1999 and 2000, the activities of members of the group spanned conferences, seminars, and informal meetings.[49]

The activities of the RMA Working Group were at least partly responsible for the spread of the notion among both military practitioners and defense scholars that Australia stood to benefit from the long-term implications of an RMA. As a former Chief of the Defense Force, General John Baker, told one audience, "Australia is one of the relatively few nations with the education, scientific, industrial, attitudinal and geographic assets to make best use of RMA possibilities."[50] Similarly, the veteran Australian strategic thinker, Professor Coral Bell, observed "the Revolution in Military Affairs offers the most promising set of systems yet evolved to solve Australia's permanent strategic dilemma: how to defend a very large territory and a long and vulnerable coastline with forces which will always remain very small by global or regional standards."[51]

In broad terms, in the period between 1999 and 2001, Project Sphinx did much to make Australian RMA thinking the most advanced in the Asia-Pacific region. Nowhere was this reality more clearly demonstrated than at a major international conference in Canberra in May 2000 entitled, "The RMA in the Asia-Pacific: Challenge and Response." The conference, initiated by the Office of the RMA and the Australian Defense Studies Center at the Australian Defense Force Academy, attracted over 200 delegates from Australasia, Europe, the Asia-Pacific, and North America. The keynote speaker was Dr. Andrew Krepinevich, the Director of the CSBA in the United States.[52]

During the proceedings, there was clear evidence, if not of a "knowledge edge" then certainly of a "knowledge gap," between Australian defense analysts and most of their Asia-Pacific counterparts. Australian speakers at the conference talked about a future battlespace environment in which network-enabled synchronized operations, tailored effects, cyber-maneuver, and joint warfighting would predominate.[53] In contrast, most Asian speakers stressed the marginal position that the RMA held in their current strategic thinking. In terms of theory, if not yet capabilities, there is little doubt that Australia has already achieved a substantial "knowledge edge" in South-East Asia. Only Singapore would appear to have any potential to match Australia in RMA thinking.[54]

DEVELOPING THE KNOWLEDGE EDGE, 1999–2002

Between the end of 1999 and the beginning of 2001, the ADO concentrated on developing the concept of a Knowledge Edge as the centerpiece of an Australian RMA. Between late 1999 and the beginning of 2001, a series of official reports, discussion papers, and briefings were produced examining the implications of an information-based military revolution. In December 2000, a Defense White Paper confirmed the concept of the Knowledge Edge as being at the heart of Australia's defense planning in the first decade of the twenty-first century. In September 2002, the Knowledge Edge entered official ADF doctrine as a key warfighting concept.

The RMA Paper, Defense Review 2000 and the Knowledge Staff

In November 1999, the ADF's Military Strategy Branch defined a revolution in military affairs as comprising "fundamental changes in the conduct of military operations resulting from innovative use of technologies, concepts and organizations in response to political, economic, security and social uncertainty."[55] Such a holistic definition placed a premium on outlining an integrated approach to an Australian RMA. As Brigadier S. H. Ayling, Director General Military Strategy put it in May 2000, "[it is] the combination of organization, doctrine and technology that leads to a superior military capability."[56]

Between late 1999 and early 2000, there was a systematic attempt to come to terms with the multidimensional demands of the RMA through the Military Strategy Branch's preparation of a major paper entitled, *The Revolution in Military Affairs and the Australian Defense Force*.[57] This official document attempted to map the direction of a distinctly Australian approach to an RMA and began the process of explaining the strategic significance of achieving a Knowledge Edge. *The Revolution in Military Affairs and the Australian Defense Force* was originally conceived for release as a public discussion paper during 2000. However, although a final version of the paper was completed and even quoted in the media, ultimately the document was not released for public debate.[58]

The official RMA paper called for a specifically Australian approach to the emerging information-based RMA.[59] Such an approach needed to be based on a judicious mixture of enabling technologies, upgraded platforms, organizational change, and new military doctrine.[60] C4ISR technologies, integrated logistics support, and information operations were identified as central to the ADF's ability to undertake effective joint and combined operations in the twenty-first century.[61] The problem of maintaining interoperability with the United States while maintaining an ability to be able to undertake independent operations in the Asia-Pacific region was also emphasized.[62]

Several of the central ideas in the Office of the RMA paper were subsequently reflected in *Defense Review 2000—Our Future Defense Force: A Public Discussion Paper*, an official publication released in June 2000.[63]

The document was published as a companion document to the work of a Community Consultation Team headed by former Foreign Minister and Ambassador to the United States, Andrew Peacock. The aim of *Defense Review 2000* was twofold. First, it was hoped the document would assist the Community Consultation Team in gauging public opinion on strategic issues at a time when, because of the deployment of elements of the ADF to East Timor, defense policy had achieved a high national profile. Second, the consultation exercise was intended to help Australian policy planners engaged in drawing up the first Defense White Paper of the twenty-first century to focus on strategic areas and budget issues that were revealed as being of public concern.

Significantly, the report of the Community Consultation Team found that "there was widespread agreement that Australia should maintain the knowledge edge in intelligence, surveillance and reconnaissance capabilities."[64] This finding coincided with a basic premises of the *Defense Review 2000* that in the twenty-first century, the Australian military would rely increasingly on two features: information technology systems—especially ISR and C2 capabilities—and the skills of highly trained military personnel.[65]

Defense Review 2000 extended ideas first mooted in ASP 97 and the Office of the RMA paper. The document suggested that the importance of information technology would grow for Australia for two reasons. First, the trend toward the modernization of military capabilities in the Asia-Pacific showed no signs of abating. The discussion paper pointed out that the numbers of various advanced combat aircraft; anti-ship missile and surface to air missile systems, and electronic warfare capacities had dramatically risen in the region during the 1990s.[66] As a result, Australia's traditional advantage in maritime and air platforms was gradually being eroded. The RAAF's 71 F/A-18 Hornet tactical fighters were gradually losing parity with the best regional air forces. Upgrades in avionics, electronic warfare, and missiles to Australia's F/A-18s and to its F-111 strike bombers along with the acquisition of AEW&C aircraft were critical to regaining air-combat parity.[67]

The emphasis on upgrades and improved avionics in *Defense Review 2000* highlighted the second reason why information technology was vital to Australia's security: most of the ADF's major air–sea platforms were facing block obsolescence between 2007 and 2020. The discussion paper pointed out that by 2015 the list of platforms at the end of their service cycle would include the RAAF's F/A-18 Hornet, the P-3C Orion maritime patrol aircraft, and C130H transport fleet; the RAN's guided missile frigates, its amphibious support and support ships; and many of the Army's wheeled vehicles. In addition, by 2020, Australia's F-111 bombers, described as "the muscle of our strike force," would have reached their life of type.[68]

In the light of the twin challenges of growing regional military capabilities and an ADF heading toward obsolescence, *Defense Review 2000* reinforced the importance of the Knowledge Edge in giving Australia "a critical military capability edge" in the future.[69] In terms of re-equipping the ADF, the paper announced that "the application of technology associated with the

'Revolution in Military Affairs' . . . may present innovative capability solutions that could yield financial savings."[70] The discussion paper suggested that an RMA-style approach to defense modernization was now vital for Australia. "Information capabilities," the document stated, "are about applying the ideas of the knowledge economy to the business of fighting wars."[71] The most critical ADF assets in the future would lie not only in the power of platforms and weapons, but also increasingly in the integration of systems and skills to produce combat effects. The document went on to observe:

> Information warfare . . . the "Revolution in Military Affairs" . . . is where our comparative advantage over potential adversaries is likely to last longest. In coming years, it will be harder for Australia to match regional numbers of platforms such as ships and aircraft, but we are well-placed to keep a lead in our ability to use what we have to the best effect.[72]

However, in order to exploit sophisticated information age capabilities, the Australian–American alliance was of fundamental importance. *Defense 2000* reaffirmed that "our alliance with the US, which leads the world in these [information capabilities] areas, is vital to giving us affordable access to this technology."[73]

Alongside the RMA content in the public discussion paper, ADF Headquarters continued to refine the concept of the Knowledge Edge as "a fundamental basis for the achievement of warfighting superiority for the ADF in the Asia Pacific Region."[74] In June 2000, a concept paper drawn up by ADF Headquarters extended the definition of the Knowledge Edge:

> *A Knowledge Edge exists when,* as a result of leveraging and exploiting information, communications and other technologies, and by the application of human cognition, reasoning and innovation, *there is a comparative advantage in those factors that influence decision making and its effective execution.*[75]

Attaining decision superiority over opponents was described as the central advantage to be gained from RMA-style technologies. The key to achieving a Knowledge Edge lay therefore in a skillful combination of command and control, information, surveillance, reconnaissance, and electronic warfare (C4ISREW) capabilities. With an infrastructure based on this suite of capabilities, Australia could eventually move toward "a 'network enabled' approach to warfighting, leveraging the connectivity between sensors, commanders and weapon systems."[76]

However, as C4ISREW capabilities provided improved connectivity in network-enabled military operations, there would have to be corresponding changes in the nontechnological areas of Knowledge Edge activity. The latter included developing suitable doctrine for joint and combined operations; reforming both military organization and military education; realigning leadership and command authority to meet information age requirements; and maintaining suitable cohesion and morale within the ADF.[77]

By mid-2001, the Department of Defense had formed a Knowledge Staff headed by a Chief Knowledge Officer (renamed Chief Information Officer in 2002), Air Vice Marshal Peter Nicholson—who, as noted earlier, was an early proponent of "knowledge dominance" forming a central feature of an Australian RMA. The central tasks of the current Knowledge Staff were to examine complex technical issues such as interoperability with allies and the coordination of simulation exercises. A Directorate of Intelligence, Surveillance, Reconnaissance, and Electronic Warfare within the Knowledge Staff has the responsibility of developing an integrated national surveillance system.[78]

The 2000 Defense White Paper and the Knowledge Edge

In December 2000, the publication of a new White Paper, *Defense 2000: Our Future Defense Force*, provided the most detailed rationale so far advanced by Canberra's strategic planners for Australia's embrace of the Knowledge Edge. In terms of the RMA, the 2000 White Paper represented the culmination of thinking that had begun in ASP 97. The new strategic blueprint reflected over three years of close analysis of both technological innovation and of the potential for revolutionary changes in the character of warfare. As one observer has noted, "the White Paper acknowledges the overriding importance of the Revolution in Military Affairs at all levels of the ADF."[79] The document contained both a general assessment of the RMA and a specific analysis of Australia's requirements from it in order to maximize the Knowledge Edge.

The main characteristics of the RMA were identified in the White Paper as a trend toward the integration of military forces for joint operations; the networking of individual systems and capabilities to achieve whole-of-force effects and multiplied combat power; and changes to military organization and doctrine.[80] As *Defense 2000* puts it:

> RMA technologies impart the ability to know more than one's adversary in relevant areas. This can result in a decisive military advantage when linked with appropriate weapons and concepts of operation. Indeed, this will probably be one of the decisive factors in warfare over the coming decades.[81]

As foreshadowed in ASP 97 and *Defense Review 2000*, the White Paper committed Australia to the development of an advanced information technology infrastructure based on major investment and cooperation with the United States.[82] Information technology, the document declared, could confer long-range precision strike using networked platforms employing stealth technology and electronic self-protection. Sensors would increase automation and remote control would help reduce personnel.[83]

The White Paper announced that the early twenty-first-century ADF would be based on a mixture of new and upgraded platforms, information, and space-based capabilities. The F/A-18 would be upgraded using stealth technology; new combat aircraft would be acquired in 2006–07 with the first fighters to enter service in 2012.[84] The RAN's Anzac frigates would receive anti-ship missile defense and a new class of three air-defense capable ships would be locally built beginning in 2005–06. Armored personnel carriers would also be upgraded but the Army would also be equipped with a new armed reconnaissance helicopter and shoulder-fired missiles.[85]

Space-based technologies such as UAVs and uninhabited combat aerial vehicles (UCAVs) were identified in *Defense 2000* as emerging systems that offered a great deal of potential for surveillance, reconnaissance, information gathering, and eventually the delivery of combat power.[86] The White Paper also announced that advances in biological procedures and nano-technology would be monitored in order "to select and acquire expertise and capability in those technologies that offer the most advantages in gaining and maintaining the knowledge edge."[87] Advanced RMA-style technology would also be applied to improve the performance of individual soldiers. In the future, the use of micro-vehicles, night vision equipment, and sophisticated navigation techniques would assist soldiers "to move faster and see further, conduct operations over 24 hours in all terrains and have vastly improved firepower at his or her fingertips."[88]

To meet the demands of twenty-first-century warfare, an organizational review of the DSTO was necessary. The organization would "undertake a fundamental review of its program of work and its structures to ensure that it is poised to take best advantage of the emerging RMA, information and other technologies."[89] The DSTO would liaise with industry in its research into guided weapons combat systems software, data management, signal processing, and C4 systems integration.[90] Australia would also pursue a cooperative project in a major UAV program (the RQ-4A Global Hawk system) with the United States and would undertake extensive research into information operations, simulation, and modeling in a series of both qualitative and quantitative wargames.[91]

To facilitate the drive toward cutting-edge RMA/Knowledge Edge research, the White Paper designated Information Capabilities to be an integral part of an $A16 billion, ten-year Defense Capability Plan (DCP) unveiled in the document. Under the DCP, Information Capabilities became, for the first time, a separate grouping in order to ensure their strategic priority.[92] Between 2001 and 2011, A$2.5 billion will be spent on developing the Information Capabilities group. Indeed, in terms of capital expenditure, information technologies now rank third in Australia's defense spending hierarchy—behind air combat (A$5.3 billion) and land forces (A$3.9 billion)—but well ahead of maritime forces (A$1.8 billion) and strike (A$0.8 billion).[93] The order of these priorities demonstrates the importance the Department of Defense now assigns to information techniques in twenty-first-century warfare.

Accordingly, there is to be sustained investment in enhanced intelligence capabilities—described in the document as critical to providing a "war-winning edge to forces in the field."[94] These capabilities include enhanced signal intelligence and imagery collection; improved geo-spatial information systems; and deeper levels of U.S.–Australian cooperation in key information systems. A specific objective is to finalize a comprehensive national surveillance system to provide continuous coverage of Australia's vast and extended northern maritime approaches. Data from the Jindalee Over-The-Horizon Radar (JORN) system would eventually be fused with other sensor systems to provide an integrated 24-hour national surveillance picture.[95]

Australia would also continue to seek to use information technology to overcome its geographic size and distance. In this respect, there is to be investment to create a networked command system to support deployed forces on operations using a single collocated theater headquarters and two deployable joint force headquarters for concurrent operations.[96] Finally, there are requirements to maximize integrated logistics systems for complex operations at short notice, provide protection against hostile information operations, and maintain a high-level interoperability with major allies.[97]

The priority afforded to the Information Capabilities grouping was justified in *Defense 2000* on two main grounds. First, the White Paper clearly viewed RMA developments as offering Australia unique advantages in information technology that were "unthinkable even a few years ago."[98] Second, the document expressed a long-held belief that embracing information technology works to a national strength since Australia possesses widespread and high-level skills in computer literacy. The combination of RMA information technologies and high computer literacy is seen by many Australian strategic planners as providing a societal base to ensure that "the 'knowledge' edge ... will be the foundation of our military capability over the coming decades."[99]

The 2000 White Paper's belief in the advantages of Australia developing information capabilities to achieve advantages in warfare was reinforced when, in September 2002, the Knowledge Edge entered Australian military doctrine. The ADF's capstone manual, *Foundations of Australian Military Doctrine*, affirms the necessity for Australia to adopt a knowledge-based approach to warfare.[100] The publication goes on to describe the Knowledge Edge as a capability that has the potential to permit Australian forces to apply combat power based on decision superiority, the operational art, and maneuver warfare principles.[101]

CHALLENGES TO THE AUSTRALIAN KNOWLEDGE EDGE: THE BUDGET CRISIS AND A CHANGING SECURITY ENVIRONMENT, 2000–03

Although Australian defense planners expect much from the long-term benefits of the Knowledge Edge and an increasingly networked military, success depends not simply on ideas and concepts, but also on adequate resources and upon the course of strategic events. Since 2000, it has become evident that if an indigenous Australian RMA is to fulfil its promise, then, problems associated with defense spending must be resolved realistically and boldly by policy-makers. Closely related to problems in Australian defense expenditure has been the drastic change to the global security environment following the al-Qa'ida attacks on the United States on September 11, 2001, and which led to military campaigns in Afghanistan and Iraq. The nexus between a lack of resources and the rise of strategic uncertainty, emanating from the war on terror, have emerged as potential obstacles to Australia's development of future warfare capabilities.

Low Defense Spending and the 2000 Budget Crisis

In 1984, Australia was spending 2.9 percent of GDP on defense. By 1999, the figure had dropped to 1.8 percent (A$11.2 billion)—the lowest percentage since 1938—representing a drop of 35 percent over 15 years.[102] By early 2000, it was clear that, unless the defense budget was substantially increased, the ADO would not be able to undertake even a modest RMA and simultaneously retain high preparedness for regional contingencies such as the peace enforcement mission in East Timor.[103]

In 1999, the leading Australian strategic analyst, Paul Dibb, predicted a "coming train smash" in Australian defense policy stemming from the Government's ambition to invest in information age capabilities being unmatched by increased defense spending. Dibb believed that the purchase of new systems and platforms along with expenditure on upgrades, enhancements, refits, and operational deployments could not be met from within a static defense budget.[104] Dibb was not alone in his concern. In April 2000, the Secretary for Defense stated bluntly that "the bottom line is that Australia can no longer afford a balanced, self-reliant, capable, and ready defense force of 50,000 with its current capabilities on 1.8% of GDP."[105] The Secretary pointed to a "convergence crisis" within Australia's defense establishment:

> The irony of our professional military performance in East Timor is that it masks the reality we face. *Australia's national security is challenged by a convergence of financial, management, planning and strategic pressures.* The Australian Defense Organization's ability to present a range of capability and military response options to Government will be severely constrained if these combined pressures are left unchecked. This crisis, which has been building over the last [post–Cold War] decade, has now come to a head due to increased personnel costs and the costs of expanding and re-equipping the capabilities of the ADF.[106]

The weakness of the defense budget was exacerbated by the problem of unreformed Cold War organizational and managerial practices. In the relatively predictable strategic environment of the later Cold War, the ADO had developed the practice of holding down operations and personnel budgets in order to fund capability and platform modernization. However, this approach to managing capital equipment and project management proved untenable in the more unpredictable conditions of the post–Cold War era. By the late 1990s, the needs of capability development and short-notice military deployments could not be met simply by resorting to scaling back spending on operational needs and personnel.[107]

By 2000, there was not enough money available to meet the triple demands of technology upgrades to existing platforms; the purchase of new platforms; and acquiring RMA/Knowledge Edge systems. With a defense expenditure base of 1.8 percent of GDP, the possibility of Australia developing both advanced high-technology military capabilities while maintaining a credible ADF for current contingencies seemed rather bleak. As Hawke bluntly put it, "at present and anticipated levels of funding, the ADF as we know it today will cease to exist."[108]

The Political Economy of Defense: The "Hi-End-Low-End" Division of 2000

The growing Australian defense budget crisis became an acute political issue during the course of 2000 and led to a fierce debate on future capabilities in the Howard Government's National Security Committee (NSC) of the Cabinet.[109] Division developed within the NSC over whether Australia required a "high-end" (shorthand for an expensive, high technology) or a "low-end" (shorthand for a cheaper, lower technology) military establishment. As one defense correspondent, Robert Garran, observed succinctly, "at the heart of the debate [in the Howard Government] is whether the Australian Defense Force should focus on peacekeeping and low-level contingencies in the region or whether it needs a powerful high-tech capability."[110]

According to various press reports, those who supported a high-end force included John Moore, the Minister for Defense and Alexander Downer, the Foreign Minister. Skeptics of the high-end force were reported to include the Treasurer, Peter Costello, the Finance Minister, John Fahey, and the influential Secretary of the Department of Prime Minister and Cabinet, Max Moore-Wilton.[111] The national daily newspaper, *The Australian*, recorded the progress of this complex, internal political debate.[112] In a series of editorials and opinion pieces, the newspaper warned against the idea that the East Timor peace enforcement experience could serve as a model for Australia's future military force structure. In January 2000, in an editorial on the implications of the RMA for Australia, *The Australian* stated, "for 'revolution in military affairs' read 'information revolution'...the attempts by the military...to deliver the capability of destroying targets with unparalleled precision." To exploit the RMA, Australia required a clear strategic approach in order to permit the ADF to "determine the best mix of [information] technologies."[113]

However, in August 2000, those favoring a low-end force and restricted defense spending in the NSC appeared to score a major victory when the government reduced the number of AEW&Cs wanted by the RAAF from seven to four aircraft. It was noted that Australia's East Timor deployment was expected to cost of over $4 billion in the period from 1999 to 2003. One low-end advocate in the cabinet was reported as prefacing his opposition to advanced warning aircraft by asking rhetorically of Defense officials: "what use would AEWCs have been in Timor?"[114] In September 2000, in yet another hardline editorial, *The Australian* warned the Government that "the capability to defend ourselves should be paramount in Cabinet thinking. It would be a national disgrace—as well as irresponsible—to argue that we can ignore the need to sustain capable military forces."[115]

The White Paper's Defense Capability Plan

By the end of 2000, it was clear that, despite tactical reverses over AEW&C capabilities, the advocates of a high-end ADF had prevailed in the political debate over defense spending in the government. In its December 2000 White Paper, the Howard government sought to provide a long-term resolution to the budget crisis. The government sought to balance strategic demands, defense capabilities, and levels of defense funding by introducing the ten-year DCP.[116] The DCP—as already noted with a strong emphasis on information technologies—was unveiled as the cornerstone of *Defense 2000*.

The aim of the DCP was to establish parameters against which defense spending could be increased by an average of about 3 percent per annum in real terms between 2001 and 2011.[117] Significantly, Prime Minister Howard declared *Defense 2000* to be "most comprehensive reappraisal of Australian defense *capability* for decades."[118] The victory of the government's high-enders was captured by *The Australian*'s banner headline on the White Paper: "Enter the cyber warriors."[119] Under *Defense 2000*'s ten-year capability plan, the Australian defense budget is to increase by A$500 million between 2000 and 2001; by A$1 billion between 2002 and 2003; and thereafter by 3 percent real growth per year until 2010. Some sources estimate that there will be an A$23.5 billion increase in expenditure in real terms over the first decade of the twenty-first century. In theory, by 2010, defense spending will stand at A$16 billion as opposed to A$11.2 billion in 2000.[120] Paul Dibb has suggested that the firm financial commitment under the DCP has made the new strategic blueprint "a benchmark Defense White Paper."[121]

However, it is important to note that the DCP remains an unbinding commitment on future Australian governments. For this reason, some observers are pessimistic about the emergence of a future high-technology ADF with a Knowledge Edge capability. As Greg Sheridan, the foreign editor of *The Australian* observed, the real cause for concern with the DCP is that "no government has ever sustained a real increase of 3 per cent in defense spending for 10 years."[122] In Sheridan's view, "the Government's White Paper is all about Australia's strategic decline. It's about managing, slowing, but above all accommodating, our national strategic decline."[123]

Defense Spending and the Implications of a Changing Security Environment: The 2003 Defense Update

Although the DCP was a serious attempt to address the budgetary problems facing Australia's defense effort, events since its publication suggest that Greg Sheridan's pessimistic view is closer to reality than that the optimism expressed by Paul Dibb. Since 2000, the prescriptive ten-year capability plan has become the victim of rapid changes in international security conditions.

By 2003, the harsh reality was that Australia's strategic environment bore little resemblance to that outlined in the 2000 Defense White Paper—a document that increasingly appears to belong to the twentieth rather than the twenty-first century.

The outmoded content of *Defense 2000* was highlighted in February 2003 when the Australian Government published its first official reaction to the changes in the global strategic environment since September 11, 2001. The new document, entitled *Australia's National Security: A Defense Update 2003*, observed that while the White Paper had "set out a Defense posture for the times"—based on the defense of geography—Australia now faced a range of different dangers from those of the 1990s.[124] The *Defense Update* identified transnational threats from weapons of mass destruction, long-range ballistic missiles and global terrorism as representing the main dangers to Australia's security in the early twenty-first century.[125] In August 2003, the Treasurer, Peter Costello, reflecting on the new security environment, threw doubt on the future of the White Paper's ten-year capability plan. He observed that because Australia's strategic circumstances had changed so drastically from the late 1990s, the measures outlined in the DCP were "not immutable."[126]

However, perhaps the most revealing insight into the nexus between the political economy of Australian defense spending and the changing security environment, came when the government-funded Australian Strategic Policy Institute (ASPI) produced a detailed report on the defense budget entitled *Sinews of War*.[127] The August 2003 ASPI report highlighted how rising costs in operations and emerging technologies, along with demands stemming from new strategic conditions, had struck at many of the assumptions of the 2000 DCP. For example, the survey reported that over the period 1999–2000 to 2004–05, deployments to East Timor, Afghanistan, and Iraq were expected to cost some A\$2.8 billion—expenditure that had been impossible to predict in the DCP.[128]

The 2003 ASPI report went on to note that over the next five years, logistics costs to maintain aging platforms in the ADF inventory, such as the F-111 strike aircraft are estimated at A\$1.1 billion.[129] Moreover, new homeland security measures have required the finding of an additional A\$500 million to fund a new Special Operations Command, an extra Special Force tactical assault group, another company of the Commando Regiment and the formation of an Incident Response Regiment to deal with chemical, biological, and nuclear crisis contingencies.[130] The result of such rising costs for operations, equipment, and personnel in the ADF has been the erosion of investment in future capabilities. For example, in 2003, A\$642 million of planned capital investment in future capabilities have been rescheduled in order to meet the demands of the ADF's current operating costs.[131]

Sinews of War also pointed out that the 2000 DCP had come under growing pressure from what the report describes as "the [problem of the] translation of Australia's emerging strategic posture into concrete decisions about capability priorities."[132] Although the 2003 *Defense Update* had highlighted non-state threats to Australia that transcend strategic geography, it refrained from indicating what changes might be required in the ADF's future force structure. The ASPI report detected uncertainty in official defense policy circles and commented:

> It is difficult to escape the conclusion that the Government has not yet decided the extent to which the events since September 11 require changes in the DCP, and indeed for defense policy more broadly. It will be hard to reach clear Defense budget decisions until these issues are resolved.[133]

In November 2003, the Australian Government released a Defence Capability Review (DCR) that sought to adjust the imperatives of the 2000 DCP to the new strategic environment outlined in the February 2003 *Defence Update*.[134] The new Review identified requirements to strengthen the Army, to enhance the ADF's strategic sea lift and to provide better air cover for deploying forces. The DCR stated that the above measures were necessary in order "to position the Australian Defence Force to exploit current and emerging Network Centric Warfare advantages".[135]

In particular, the DCR announced that the Army would be re-equipped with new main battle tanks for improved combined arms warfare. The new tank will be chosen in 2004 from three contenders—the US Abrams, the German Leopard, and the British Challenger.[136] The RAN's projected air warfare destroyers would be fitted with a variant of the US Aegis air warfare system and larger amphibious ships would be acquired for increased strategic lift. In terms of air capability, the Review confirmed that the RAAF would acquire up to 100 F-35 Joint Strike Fighters as its replacement combat aircraft. Upgrades to the F/A-18 fighters would proceed while new AEW&C aircraft and Global Hawk UAVs would also be purchased. Finally, the aging F-111 strike bomber fleet is to be withdrawn from service by 2010, rather than by 2020, as envisaged in the original 2000 Defense Capability Plan.[137] With its emphasis on purchasing American equipment—the Abrams tank, a modifed Aegis air warfare system, Global Hawk UAVs, and the Joint Strike Fighter—the DCR appears to reinforce the ADF's capacity for undertaking interoperable coalition operations within the context of a strengthened Australian–American security relationship.[138]

TOWARD NETWORK-CENTRIC WARFARE, 2002–04

Despite budgetary difficulties and the impact of a rapidly changing strategic landscape, Australia's conceptual development of the Knowledge Edge continued. During 2002, the Office of the RMA and Project Sphinx were absorbed into a new Directorate of Future Warfare and Concepts inside the ADF's Military Strategy Branch.[139] The ADF also began to focus more directly on developing network-enabled operations, an approach first used by the Australian Army in the late 1990s and subsequently adopted by the Knowledge Staff in mid-2000. In 2002, two new ADF booklets, *Force 2020* and *Future Warfighting Concept*, brought network-centric warfare to the forefront of Australian theorizing about the RMA and future warfare.[140]

Force 2020 and Network-Enabled Operations

In June 2002, when the Military Strategy Branch published *Force 2020*, the intention was to provide the ADF with a "vision statement" that would permit defense planners to translate the Knowledge Edge from the idiom of theory into the arena of practice. The new booklet introduced three new ideas to Australian RMA planning—the notion of a seamless force and the twin concepts of networked-enabled operations and effects-based operations.[141]

Force 2020 envisaged the future transformation of the ADF from a joint into a "seamlessly integrated force" in which "platform-centric" operations gradually give way to what the publication calls "network-enabled operations."[142] In a seamless and integrated ADF, platforms would increasingly be treated as "nodes" in a network. Such "nodes" would be capable of collecting, sharing, and accessing common information across the battlespace to increase combat power. The fundamental building block of a future integrated system would be the creation of a comprehensive "information network" composed of a tiered system of grids. These grids include a *sensor grid* (collating real-time information as a primary source of combat power); *a command and control grid* (to collate and analyze information) and an *engagement grid* (executing precision engagement by directing "shooters" at targets).[143]

It is important to note that *Force 2020's* vision of a networked approach to warfare is closely linked to the idea of effects-based operations.[144] The authors of *Force 2020* note that an effects-based approach, from the national level, is a preferred approach "because it meets the emergent breadth of *threat*—not just military challenges, but also broader security challenges—and provides government with most options."[145] By focusing on effects, or the outcomes from applying military power, strategy becomes more subtle, inter-agency, and multidimensional in character.

Force 2020 also reaffirms what it describes as a "concept-led Long-Range Planning" methodology based on experimentation and simulation—an approach that has underpinned the Australian attitude toward future warfare since the mid-1990s.[146] In this area of activity, the work of the Military Strategy Branch remains closely linked to the future warfare analysis being undertaken by various single service experimental teams. The work of these three teams, *Headmark* (the RAN Future Maritime Operational Concept) *Headline* (the Army's Experimental Force based on Maneuver Operations in a Littoral Environment), and *Headway* (the RAAF's Future Aerospace Warfare Concept) remains ongoing and informs the ADF-wide effort.[147]

Future Warfighting, Network-Centric Warfare and the Concept of Multidimensional Maneuver

In December 2002, *Force 2020* was followed by the publication of the ADF's *Future Warfighting Concept*, the product of three-years work by the Military Strategy Branch.[148] The publication enlarged on the three major innovations of *Force 2020*—the seamless force, network-enabled operations (now renamed network-centric warfare) and effects-based operations—and examined each in the context of fluid and unpredictable security conditions.[149] In many respects, the new booklet provided an Australian perspective on changes to the international security environment since September 11, 2001.

Future Warfighting Concept is not, however, a policy document, but rather a meditation on Australian aspirations in future warfare. It represents a "stake in the ground" that seeks to guide and synchronize both joint- and single-service concept development. The booklet attempts to provide a common understanding within the ADF of the vital role that concepts and experimentation will assume in future ADF capability development. As the new publication puts it, "[this booklet] presents a concept that describes how we aspire to fight in the future, and a way to develop new ideas about future capability."[150] The publication outlines the challenge of future warfare to the ADF in the following terms:

> Ideas such as network-centricity, an effects-based approach, taking a systems view of the adversary and ourselves, and concept development and experimentation are only beginning to emerge in the ADF's thinking about conflict...We should not underestimate the degree of change that these ideas will bring... We expect to need ten to fifteen years to realise the FWC [Future Warfighting Concept].[151]

Future Warfighting Concept is divided into two parts. The first part of the publication describes the factors that influence the ADF view of joint warfighting and attempts to identify warfighting requirements under conditions of globalized security involving the rise of non-state actors, nonlinear warfare, and the blurring of "conventional" and "unconventional" aspects of conflict. The booklet argues that the emerging extraterritorial battlespace of the new millennium has broadened the spectrum of conflict to encompass maritime, aerospace, land, electromagnetic, and temporal, social and political dimensions of conflict.[152] These new developments challenge Australia's late twentieth-century mode of strategic planning based on the principles of strategic geography:

> Technology will...influence the view of geography in our security planning. We have already seen how international terrorism, using a mix of technologies... can circumvent borders and distance to attack our national interests. Other technological developments such as offensive information capabilities, space-based sensors and communications, weapons of mass effect, and long-range weapons such as ballistic and cruise missiles have the potential to reach targets that were previously difficult to strike.[153]

Building on the ideas first outlined in *Force 2020* in June 2002, *Future Warfighting Concept* describes the emergence of a national effects-based approach to strategy based on "the application of military and non-military capabilities to realize specific and desired strategic and operational outcomes in peace, tension, conflict and post-conflict situations."[154] An effects-based strategy is defined as aiming to promote a "whole-of-nation view of security" while "effects are the physical, functional or psychological outcome, event or consequences that results from specific military or non-military actions at the tactical, operational and strategic level."[155]

The second part of *Future Warfighting Concept* explores the use of network-centric warfare in facilitating an effects-based strategy. Network-centric warfare is described as offering a range of warfighting advantages, including the ability to focus superior knowledge and increased protection for Australian forces in their operational activities.[156] More specifically, the ADF's aim is to use a networked military to execute its future warfighting concept of "multidimensional maneuver." Network-centric warfare is described as "a key enabler that will allow us to conduct multidimensional maneuver, and achieve the seamless force envisaged in *Force 2020*."[157]

The concept of multidimensional maneuver appears to be based upon the notion that Australia must move beyond a mechanistic attrition-model of continental defense, toward a new approach to war in which military forces exploit network-centricity and effects-based operations in a systems-approach to warfare.[158] Like most joint warfighting concepts, the ADF's construct of multidimensional maneuver relies upon a synthesis of single-service ideas adapted to the arena of inter-service cooperation. The ADF seeks to harmonize single-service visions of future land, naval, and aerospace warfare and to subject them to the strategic-level discipline of a national effects-based approach employing systems thinking and net assessment.[159]

It is not surprising then that the new future warfighting concept contains a rather eclectic mix of ideas. Not all of these ideas have yet been reconciled or properly defined in Australian military doctrine. In the ADF's concept of multidimensional maneuver, it is possible to detect the influence of Liddell Hart's theory of the strategy of indirect approach (promoted by the Army); John Boyd's idea of exploiting the competitive decision cycle (promoted by all three services); and John Warden's theories on strategic paralysis (much favored by the RAAF).[160] Moreover, the general intellectual framework of an effects-based strategy in which these ideas have been situated, seems to have been at least partly influenced by a military interpretation of Ludwig von Bertanlanffly's theory of general systems. Under systems theory, an adversary is not seen as a disconnected array of tactical targets to be destroyed by attrition, but rather as a system of interacting components that can be disrupted by the strategic effects of operational maneuver.[161]

The ADF and American and Swedish Approaches to Network-Centric Warfare

There have also been more immediate influences at work in Australian ideas about future warfare. The most notable influences on the ADF's approach to networked warfare over the past three years has been the U.S. Office of Force Transformation headed by Vice-Admiral Arthur K. Cebrowski and the Swedish Armed Forces' work on network-based defense. The involvement of Cebrowski's organization in Australian future warfare can be seen as a natural outgrowth of previous Australian–U.S. links—notably between the former Office of the RMA and Andrew Marshall's ONA and Andrew Krepinevich's CSBA—during the 1990s.

Admiral Cebrowski's various writings, propounding a historic shift from platforms to networks and his belief that network-centric operations amount to a "new theory of war," have long been informally influential in official strategic circles in Canberra.[162] The creation of the U.S. Office of Force Transformation under Cebrowski's leadership in 2001 led to the establishment of more formal links between Military Strategy Branch and the new American organization. The influence of Cebrowski's ideas in Australia's approach to networked-warfare was evident in May 2003 when John J. Gartska, Assistant Director for Concepts and Operations in the Office of Force Transformation, delivered a major address at the ADF's inaugural network-centric warfare conference held in Canberra.[163]

The role of the Swedes in the development of Australia's networked-approach to future warfare is more recent. On the surface, Sweden— a European country with few natural ties to Australia and one that last experienced war in the age of Napoleon—would appear to be an unlikely RMA model for Canberra. Yet, as a small, but technologically advanced defense force adapting from Cold War geographically based strategic posture toward twenty-first-century networked-warfare, the Swedish armed forces provide lessons for Australia to study. Other parallels between Australia and Sweden include exactly the same level of expenditure on defense (1.9 percent of GDP in 2002) and a degree of similar status as "middle powers" in important strategic regions—Sweden in Europe and Australia in Asia.[164]

Above all, however, Sweden's *Natverks Baserat Forsvars* (Network Based Defense) policy, first authorized by the Stockholm Parliament in 2000, has provided the ADF with a compact contrast to the vast undertaking of America's military transformation. As a result, there has been a regular exchange of ideas between the Swedish military and the ADF over the past two years. These links culminated in attendance by Swedish military representatives at the inaugural ADF network-centric warfare conference held in Canberra in May 2003.[165]

Of particular interest to Australian military planners has been the architecture behind Sweden's Ledsyst Project or "roadmap system." The latter is a ten-year plan to create a network-based defense system combining joint command and control, a sense and weapons grid and stealth technologies and precision strike with a focus on interoperability with allies. The Ledsyst Project involves three phases: a study phase (2000–02), a demonstration phase (2002–07), and an acquisition phase (2007–10). This phased, component-based approach to the creation of a Swedish "system of systems" has been compared to assembling the building pieces in a child's Lego set.[166]

Overall, like Australia, Sweden's approach to the RMA involves a move away from massing forces in platform-oriented operations toward massing the effects of those platforms and other systems through the cumulative power of a network-based military.[167] Over the next four years, the Swedish armed forces plan to develop networking demonstrations for battlespace awareness in key areas such as cruise missile defense, territorial defense, urban warfare, and maneuver operations.[168]

The ADO's May 2003 network-centric warfare conference in Canberra provided a catalyst for the integration of Australian, American, and Swedish ideas about the potential of network-centric warfare. The conference signified that a consensus had developed in Australian defense circles over the likely benefits to be gained from networking military forces and capabilities. As the Minister for Defense, Senator Hill put it, "Network Centric Warfare is recognised as a key element in maximising our military effectiveness."[169]

The conference concentrated on developing a "roadmap strategy" for the ADF to develop practical measures in networking its military forces. At the gathering, some aspects of the Swedish Ledsyst Project provided examples for the ADF's "roadmap strategy"—with the latter aimed at creating a "single virtual network" over the next 20 years.[170] There are three objectives in the ADF's "roadmap strategy." The first is to achieve joint interoperability (the seamless, integrated force). The second is to develop interoperable force options for use within a U.S.-led international coalition (such as Afghanistan or Iraq). Third, the ADF seeks the capacity to operate with, and potentially lead, coalitions within military partners in the area of its immediate interest, the Asia-Pacific region.[171]

The May 2003 conference also saw the launch of a draft ADF discussion paper, "Enabling Multidimensional Maneuver" outlining Australian aspirations in network-centric warfare over the next two decade or more.[172] According to the discussion paper, Australia's roadmap methodology is designed to proceed by "learning by doing" based on selective modernization of key defense programs. These programs include joint offensive support, maritime surveillance, force logistics, and air defense. The aim is to integrate such programs into functional networks to facilitate "access to the right information at the right time, allowing the ADF to place the right forces in the right place to achieve the right effect."[173]

Over time, all ADF elements are envisaged as becoming part of a "single virtual network" in which information is passed to military components through a web of interlinked sensor, engagement, and command grids. By 2015, it is hoped that the ADF will use a "system of systems" employing advanced data links that can be assembled into a common operational picture.[174] The 2003 discussion paper, like that of the earlier 1999 RMA paper, continues the pragmatic approach adopted by Australian planners toward information-based future warfare. For example, the 2003 discussion paper admits that network-centric warfare is an unproven doctrine that is vulnerable to the laws of physics, the friction of war and to a variety of deception techniques.[175] A transparent battlespace delivering perfect omniscience is, the document concludes, unlikely to emerge. The paper observes:

> The [network-centric warfare] concept has its weaknesses. We do not have a complete picture of the types of doctrinal and organizational change that are

required...Nor do we fully understand how people will manage the vastly
increased amount of information [in warfighting]...Just as some markets were
overawed by the "new economy" and neglected the facts of economic life,
some proponents of NCW are overawed by information and forget that warfare is ultimately about fighting.[176]

Despite these concerns, Australia remains firmly committed to developing a
functional approach to network-centric warfare. A recent "lessons learned"
study of the ADF's role in the invasion of Iraq in 2003 highlights how networked military operations were "crucial success factors" during the war to
disarm Saddam Hussein.[177]

CONCLUSION

Over the last decade, the Australian RMA initiative has moved from an informal, service-driven debate about "knowledge dominance" through to the
official formulation of a Knowledge Edge and associated information capabilities, to the formal adoption of network-centric warfare. In the realm of
RMA ideas, Australia has borrowed liberally from the United States experience and, more recently, from Sweden, while always remaining capable of
indigenous theorizing for its specific needs in future warfare.

The main characteristic of the Australian Knowledge Edge approach to
the RMA is its pragmatism, its focus on warfighting and its relatively modest
set of goals. In this respect, Australian planners have conformed to
C. P. Snow's famous 1961 warning that, for official technological research to
succeed in a democracy, its objectives must be both clear and "not too
grandiloquently vast."[178] In an intellectual sense, Australian defense planners
have generally avoided viewing information-based warfare as a utopian technological ideology. They have eschewed what one scholar has styled as a
"cockroach-exterminator model of warfare" based on the mathematics of precision-strike weapons.[179] Nor have Australian planners embraced network-
centric warfare in terms that are similar to Admiral Cebrowski's "new theory
of war." In short, the Australian approach to war, while valuing technology,
remains focused on the importance of Clausewitzian political and human
dimensions in armed conflict.

The greatest weakness in Australia's application of information technology
to the future battlespace remains undoubtedly one of financial resources.
This weakness has been further exacerbated by the reality that the
Government's solution to the budgetary problems—the much vaunted 2000
DCP—has been overtaken by military events. In many respects, at the end of
2003, Australia's strategic policy is at a crossroads. Old certainties have
confronted new uncertainties and the twentieth-century model of interstate
war has been supplemented by new non-state and transnational threats,
based on both asymmetric methods and the proliferation of destructive
weapons technology.

Given these new challenges, it is now clear that the 2000 Defense White
Paper represented comfortable continuity with the past rather than dynamic

thinking about the future. As a result, *Defense 2000* proved irrelevant as a guide to deal with the crises of September 11, the war on terror and the ensuing campaigns in Afghanistan and Iraq. Yet, *Defense 2000*'s interim successor, the 2003 *Defense Update*—while diagnosing the arrival of a new strategic environment characterized by global terrorism, weapons of mass destruction, and ballistic missile proliferation—provided little new guidance on Australia's future strategic direction, force structure, or capability development.

The Australian RMA initiative, launched in the relatively predictable conditions of the mid-1990s, must now learn to adapt itself to a new millennium in which peace, crisis, and war can no longer be easily separated. In an era where John Moore's regional "sea of instability" increasingly collides with the unpredictable dangers of Robert Hill's global "arc of terrorism" Australian defense planners should expect the unexpected. Ultimately, Australia's quest for a Knowledge Edge must remain firmly focused on ensuring that theory supports practice for, as Clausewitz reminds us, "the example is the real-life case, the formula the abstraction."[180]

NOTES

Dr. Michael Evans is Head of the Australian Army's Land Warfare Studies at the Royal Military College of Australia, Duntroon, Canberra.

 1. The views expressed in this chapter are those of the author and should not be seen as official representations of the Australian Army or of the Australian Department of Defense.
 2. Eighty-eight Australians were killed in the Bali bombing.
 3. Speech by the Hon. John Moore, Minister for Defense, December 6, 2001 as quoted in *The Australian*, December 7, 2000.
 4. Ben Reilly, "The Africanization of the South Pacific," *Australian Journal of International Affairs*, 54 : 3 (November 2000) pp. 261–269.
 5. Senator the Hon Robert Hill, Minister for Defense, "Regional Terrorism, Global Security and the Defense of Australia," Speech to the RUSI Triennial International Seminar, National Convention Centre, Canberra, October 9, 2003, p. 6, at <http://www.minister.defence.gov.au.> See also Geoffrey Barker, "Hill sees 'arc of terrorism'," *Australian Financial Review*, October 10, 2003.
 6. See Michael Evans, "The Middle Way: Australia's Response to the Revolution in Military Affairs," *National Security Studies Quarterly*, 6 : 1 (Winter 2000) pp. 1–19; *Australia and the Revolution in Military Affairs: The Challenge for a Middle Power*, Working Paper, Pentagon Study Group on Japan and NE Asia, Japan Information Access Project, Washington, D.C., July 24, 2000; *Australia and the Revolution in Military Affairs*, Land Warfare Studies Working Paper No. 115, Canberra, August 2001 and "Australia and the Quest for the Knowledge Edge," *Joint Force Quarterly*, no. 2 (Spring 2002) pp. 41–51.
 7. Ibid.
 8. Evans, "The Middle Way: Australia's Response to the Revolution in Military Affairs," pp. 2–8.
 9. Air Vice Marshal P. G. Nicholson. "Operating the RAAF Beyond 2000," in Alan Stephens, ed., *New Era Security: The RAAF in the Next Twenty-Five Years* (Canberra: Air Power Studies Centre, 1996) pp. 249–264.

10. Richard Brabin-Smith, "The Impact of Emerging Technologies," in J. Mohan Malik, ed., *The Future Battlefield* (Melbourne: Deakin University Press in association with the Directorate of Army Research and Analysis, 1997), pp. 139–150.

11. Dr. Jason Scholz, "DSTO and the Australian RMA Initiative," Presentation at the Australian Defense Organization RMA Seminar, November 8–9, 1999. Copy in author's possession. See also Evans, "The Middle Way: Australia's Response to the Revolution in Military Affairs," p. 5.

12. Evans, "The Middle Way: Australia's Response to the Revolution in Military Affairs," pp. 6–7 and Lieutenant Colonel G. T. Peterson, "The Impact of the Revolution in Military Affairs on the Australian Defense Force," *Yolla: Journal of the Joint Services Staff College Association*, 4 : 1 (October 1996), fn 16.

13. The Hon. I. M. McLachlan, "Defence Challenges in New Era Security," in Stephens, ed., *New Era Security*, pp. 3–8.

14. Ibid., p. 4.

15. Ibid., pp. 4–5.

16. For the impact of American ideas, see the proceedings of the first Australian RMA conference in Keith Thomas, ed., *The Revolution in Military Affairs: Warfare in the Information Age* (Canberra: Australian Defence Studies Centre, 1997).

17. Andrew Marshall, "Introduction," in Keith Thomas, *The Revolution in Military Affairs*, pp. 3–5.

18. Commonwealth of Australia. *Australia's Strategic Policy 1997* (Canberra: Directorate of Publishing and Visual Communications, 1997).

19. Ibid., p. 55.

20. Ibid.

21. Ibid.

22. For the development of Nicholson's ideas on knowledge dominance, see Air Vice Marshal Peter Nicholson, *Controlling Australia's Information Environment or Decision Superiority and War-Fighting*, Paper No. 65 (Canberra: Air Power Studies Centre, June 1998). The DSTO's Electronics and Surveillance Research Laboratory also carried out important work on the Knowledge Edge in 1996 and 1997.

23. *Australia's Strategic Policy 1997*, p. 56.

24. Ibid.

25. Ibid., pp. 56–60.

26. Ibid., p. 57. For an analysis of the implications of the Knowledge Edge, see Paul Dibb, "The Relevance of the Knowledge Edge," *Australian Defence Force Journal*, no. 134 (January/February 1999) pp. 37–48.

27. Australian Defense Headquarters, Military Strategy Branch (Office of the RMA), Minute, "Public Discussion Paper—'The Revolution in Military Affairs and the Australian Defence Force',," September 16, 1999, p. 1.

28. Evans, "The Middle Way: Australia's Response to the Revolution in Military Affairs," pp. 11–12.

29. The first Head of the ORMA was Brigadier S. H. Ayling. In June 2001, he was succeeded by Air Commodore John N. Blackburn.

30. Brigadier S. H. Ayling, Office of the RMA, "Foreword", *Australian Defence Force Journal*, no. 144 (September/October 2000), p. 2.

31. Australian Defense Headquarters. Military Strategy Branch (Office of the RMA) Minute, "Public Discussion Paper—'The Revolution in Military Affairs and the Australian Defence Force,'" September 16, 1999.

32. For background to the services' futures directorates, see Air Commodore John N. Blackburn, AM, Director General Policy and Planning—Air Force, Commodore Lee Cordner, Director General Navy Strategic Policy and Futures, and Brigadier Michael A. Swan, " 'Not the Size of the Dog in the Fight': RMA—The ADF Application," *Australian Defence Force Journal*, no. 144 (September/October 2000) pp. 65–69.

33. See Australian Army, *Land Warfare Doctrine 1, The Fundamentals of Land Warfare* (Sydney: Combined Arms Training and Development Centre, 1999), chapter 6; *The Army Continuous Modernisation Plan, 1999–2004.* Draft as at July 12, 1999, pp. 8–17.

34. Blackburn, Cordner, and Swan, " 'Not the Size of the Dog in the Fight': RMA—the ADF Application," p. 68.

35. Ibid.

36. Australian Defense Headquarters. Strategic Policy and Plans Division, "Project Sphinx," Briefing Paper by Air Vice Marshal P. G. Nicholson, Head, Strategic Policy and Plans Division, April 7, 1999. Document in author's possession.

37. Ibid., p. 1.

38. Australian Defense Headquarters, Vice Chief of the Defense Force, "Capability Executive Meeting 10 December 1999: Outcomes," pp. 2–3. Document in author's possession.

39. Brigadier S. H. Ayling, "Future Warfare Concepts: Designing the Future Defence Force," *Australian Defence Force Journal*, no. 144 (September/October 2000), p. 6.

40. Ibid.

41. Ibid., pp. 6–7.

42. Author's notes at Australian Department of Defense RMA Working Group meeting, June 30, 2000. The MSEB was to become operational in 2001.

43. Department of Defense. Australian Defense Headquarters, Brief for HSPP, "Project Sphinx: Concept Initiation Teams," May 3, 1999 and "Concept for the Krait Series of Wargames," June 1999; Ayling, "Future Warfare Concepts: Designing the Future Defence Force," p. 7.

44. Department of Defense. Directorate of Future Warfare Discussion Paper No. 1, "Project Sphinx: Military Challenges and Warfare Concepts for the ADF in 2025," no date but clearly early 1999, pp. 1–8. Document in author's possession.

45. "Concept for the Krait Series of Future Wargames," p. 2.

46. Department of Defense, Military Strategy Branch, "Terms of Reference: ADF Liaison Officer Placement with USJFCOM," January 2000, pp. 1–3; Australian Liaison Officer U.S. Joint Forces Command. Minutes, "Weekly Activity Report, Appendix: Concept Summary: Effects-Based Operations," July 26, 2000. Documents in author's possession.

47. "Concept for Krait Series of Future Wargames," pp. 6–7.

48. Ayling, "Future Warfare Concepts: Designing the Future Defence Force," p. 8.

49. Ibid.

50. General John Baker, AC, DSM, (Rtd.), "Australia's Defence Posture," *Australian Defence Force Journal*, no. 143 (July/August 2000), p. 16

51. Coral Bell, "Security Regionalisation and the Future of the Australian Defence Forces," *Australian Defence Force Journal*, no. 143 (July/August 2000), p. 21.

52. The proceedings of the conference are contained in a special edition of *Australian Defence Force Journal*, no. 144 (September/October 2000).

53. Blackburn, Cordner, and Swan, " 'Not the Size of the Dog in the Fight': RMA—The ADF Application," pp. 65–69 and Jason B. Scholtz, "Networked-Enabled Force Synchronisation," in *Australian Defence Force Journal*, no. 144 (September/October 2000) pp. 70–77.
54. Lieutenant-Colonel Hugh Lim, Singapore Armed Forces, "Impact of RMA on Command and Control—An SAF Perspective," *Australian Defence Force Journal*, no. 144 (September/October 2000) pp. 21–26.
55. Brigadier S. H. Ayling, "The Implications of the Revolution in Military Affairs." Presentation at the Australian Defense Organization RMA Seminar, Russell Offices, Canberra, November 8, 1999. Copy in author's possession.
56. Ayling, "Future Warfare Concepts: Designing the Future Defense Force," p. 6.
57. Department of Defense, Military Strategy Branch, *The Revolution in Military Affairs and the Australian Defence Force: A Public Discussion Paper* (Second Draft), December 1999. Copy in author's possession.
58. See, for example, the editorial in *The Australian* newspaper, January 4, 2000 on the discussion paper. The document was sidelined by the Government's decision in early 2000 to proceed with a broader public discussion paper on the future of Australian defense policy that incorporated aspects of the RMA.
59. *The Revolution in Military Affairs and the Australian Defence Force: A Public Discussion Paper*, pp. 1–2; 4–1.
60. Ibid., pp. 1–2.
61. Ibid., pp. 3-1–3-7.
62. Ibid., pp. 4–8.
63. Department of Defense, *Defence Review 2000—Our Future Defence Force: A Public Discussion Paper* (Canberra: Defence Publishing Service, June 2000).
64. Department of Defense, *Australian Perspectives on Defence: Report of the Community Consultation Team*, September 2000, pp. 6–7. The Community Consultation Team held 28 public meetings, surveyed 2000 people and accepted 1,157 submissions.
65. *Defence Review 2000*, p. 46.
66. Ibid., p. 14. For good analyses of Asian military modernization see Paul Dibb, "Defence Force Modernisation in Asia: Towards 2000 and Beyond," *Contemporary Southeast Asia*, 18:4 (March 1997) pp. 347–60 and "The Revolution in Military Affairs and Asian Security," *Survival* 39: 4 (Winter 1997–98) pp. 93–116. Dibb views Singapore, China, and Japan as the only three Asian states with a significant RMA potential.
67. *Defence Review 2000*, pp. 36–39.
68. Ibid., pp. 36, 37–39, 54.
69. *Defence Review 2000*, p. 46.
70. Ibid., p. 54.
71. Ibid., pp. x; 47.
72. Ibid., p. 46.
73. Ibid.
74. Department of Defense, Australian Defence Headquarters, "Brief for Vice Chief of the Defence Force on the Knowledge Edge," June 22, 2000, p. 1. Document in author's possession.
75. Ibid., p. 1. Emphases in original.
76. Ibid., p. 2.
77. Ibid., pp. 3–4.
78. Department of Defense, Australian Defense Headquarters, "The Knowledge Staff," May 23, 2001. Document in author's possession. See also Gregor

Ferguson, "JP 129 Awaits Surveillance Report," *Australian Defence Magazine* 9 : 3 (March 2001) pp. 33–34.

79. Gregor Ferguson. "Army Blessed by White Paper," *Australian Defence Magazine* 9: 3 (March 2001), p. 38. On the central importance of the RMA in the White Paper, see also Geoffrey Barker, "Defence gets a shot of realism," *The Australian Financial Review*, December 7, 2000.

80. Ibid., p. 108.

81. Ibid., pp. 108–109.

82. Ibid., p. 109.

83. Ibid., pp. 109–110.

84. *Defence Review 2000*, p. 97. The Review did not specify a replacement combat aircraft for the RAAF. However, in June 2002, Australia entered an agreement with the United States to collaborate in the development of the F-35 Joint Strike Fighter with the aim of purchasing 100 of these aircraft by 2012. The Joint Strike Fighter will replace the RAAF's F/A-18 fighters and F-111 strike bombers at a cost of A$12 billion. See Adrian McGregor, "Ready for Landing," *The Australian*, August 6, 2003 and Gregor Ferguson, "More JSF Work for Australian Firms," *Australian Defence Magazine*, February 2004, Vol. 12, no. 2. pp. 28–29.

85. See Kevan Wolfe, "Australia's Defence White Paper—An Overview," *Asia-Pacific Defence Reporter* 26: 7 (December/January 2001) pp. 21–22.

86. *Defence 2000*, pp. 109–111.

87. Ibid., p. 111.

88. Ibid., p. 111. In East Timor, Australian troops found that night vision equipment was a key factor in gaining dominance over Indonesian-backed militia forces.

89. *Defence 2000*, pp. 112–113.

90. Ibid., pp. 99–100.

91. Ibid., pp. 111–112.

92. Ibid., pp. 57, 94.

93. Ibid., p. 97.

94. Ibid., pp. 95–96.

95. Ibid., p. 96.

96. Ibid., pp. 96–97.

97. Ibid., pp. 95–96, 107.

98. Ibid., p. 94.

99. Ibid.

100. Department of Defense, *Foundations of Australian Military Doctrine* (Canberra: Australian Defence Doctrine Publication, Defence Publishing Service, September 2002) pp. 5-5–5-7.

101. Ibid., p. 5–7.

102. *The Revolution in Military Affairs and the Australian Defence Force: A Public Discussion Paper*, pp. 5–8. In 1985, the Australian Defense organization consisted of 70,000 uniformed personnel and 40,000 civilians. By 2000, these numbers had fallen to 50,000 and 16,000, respectively, *The Australian*, February 23, 2000.

103. Evans, "The Middle Way: Australia's Response to the Revolution in Military Affairs," pp. 14–15 and *Australia and the Revolution in Military Affairs: The Challenge for a Middle Power*, pp. 15–20.

104. See Paul Dibb's views as quoted in *The Australian*, September 19, 2000. Of A$11.2 billion, $6.52 billion was spent on current capability, 3.3 billion on

future capabilty, 220 million on research, 840 million on personnel services and 375 million on resource administration, *Defence Review 2000*, p. 51.

105. Department of Defense, "Address by Dr Allan Hawke, Secretary of Defence to the Royal United Services Institute for Defence Studies of Victoria," April 27, 2000, pp. 1–3. Document in author's possession.

106. Ibid., p. 2. Emphasis added.

107. Ibid., pp. 4–5; Evans, *Australia and the Revolution in Military Affairs: Challenge for a Middle Power*, pp. 15–18.

108. "Address by Dr Alan Hawke, Secretary of Defence, to the Royal United Services for Defence Studies of Victoria," April 22, 2000, p. 10.

109. For a useful overview see Graeme Dobell, "The Politics of the White Paper," *Australian Defence Force Journal*, no. 147 (March/April 2001) pp. 31–33.

110. Robert Garran, "Defence Splits Cabinet," *The Australian*, September 19, 2000.

111. Ibid.

112. See especially, Paul Kelly, "All quiet on the spending front," *The Australian*, February 23, 2000.

113. "Revolution Raises Policy Questions," *The Australian*, January 4, 2000, editorial.

114. Robert Garran, "Defence's Strategy for Survival," *The Australian*, October 10, 2000.

115. "Defence Lacks Coherent Strategic Lead," *The Australian*, September 21, 2000, editorial.

116. *Defence 2000*, pp. vii, 6.

117. Ibid., pp. 117.

118. Statement by Prime Minister John Howard, *The Sydney Morning Herald*, December 7, 2000.

119. *The Australian*, December 7, 2000.

120. See Wolfe, "Australia's Defence White Paper—An Overview," pp. 22–24.

121. Paul Dibb, "Defence Paper is Evolutionary," *Australian Financial Review*, December 7, 2000 and "Australia's Best Defence White Paper?," *Australian Defence Force Journal*, no. 147 (March/April 2001) pp. 27–8.

122. Greg Sheridan, "New equipment aside, post-coital blues are inevitable," *The Australian*, December 1, 2000.

123. Greg Sheridan, "Chronicle of strategic decline," *The Australian*, December 7, 2000.

124. Department of Defense, *Australia's National Security: A Defence Update 2003* (Canberra: Commonwealth of Australia, 2003) pp. 23–4.

125. Ibid., pp. 8–17.

126. Ian McPhedran, "Defence blueprint attacked," *Herald Sun* (Sydney), August 21, 2003.

127. Australian Strategic Policy Institute, *Sinews of War: The Defence Budget in 2003 and How We Got There* (Canberra: Australian Strategic Policy Institute, August 2003).

128. Ibid., p. 23.

129. Ibid., pp. 21–22.

130. Ibid., p. 19.

131. Ibid., pp. 23–24.

132. Ibid., p. 26

133. Ibid., p. 27.
134. Media Release, Senator The Hon Robert Hill, Minister for Defence, "Statement: Defence Capability Review," 142/2003, November 7, 2003.
135. Ibid. p. 2.
136. Ibid. In a follow up to the November 2003 DCR, in March 2004, the Australian Government announced that it had decided that the Army would purchase 59 U.S. built M-1A1 Abrams tanks at a cost of A$550 million. See Peter Jennings, "Tanks Give Troops Vital Support," *The Australian* March 11, 2004.
137. Statement: Defence Capability Review, November 7, 2003, pp. 2–4.
138. See Rod Lyon and William T. Tow, *The Future of the Australian–US Security Relationship*, Strategic Studies Institute, U.S. Army War College, Carlisle, PA, December 2003, pp. 20–35.
139. The Military Strategy Branch consists of four Directorates: Military Strategy and Strategic Doctrine; Future Warfare and Concepts; Strategic Wargaming; and Force Structure Guidance. Project Sphinx became known as the Sphinx Discussion Forum.
140. Department of Defense, *Force 2020*, Public Affairs and Corporate Communication, Canberra, June 2002 and *Future Warfighting Concept*, Canberra (Policy Guidance and Analysis Division, December 2002).
141. *Force 2020*, pp. 17–22.
142. Ibid., pp. 15–18.
143. Ibid., p. 19.
144. Ibid., p. 22.
145. Ibid. Emphasis in original.
146. Ibid., p. 25
147. Ibid., p. 19
148. Department of Defense, *Future Warfighting Concept* (Canberra: Policy Guidance and Analysis Division, December 2002).
149. Ibid., p. 2.
150. Ibid.
151. Ibid., p. 3.
152. Ibid., 8.
153. Ibid., p. 9.
154. Ibid., p. 12.
155. Ibid., p. 11.
156. Ibid., p. 29.
157. Ibid., p. 29.
158. Ibid., pp. 2–3.
159. Ibid., p. 12.
160. The author was involved in discussions in 2001–02 between the ADF's Military Strategy Branch and the three service think tanks, the Army's Land Warfare Studies Center, the RAN's Seapower Centre and the RAAF's Aerospace Centre, on developing the future warfighting concept.
161. For a discussion of the military application of systems theory, see Shimon Naveh, *In Pursuit of Military Excellence: The Evolution of Operational Theory*, (London: Frank Cass, 1997) pp. 4–29.
162. See Vice-Admiral Arthur K. Cebrowski and John H. Gartska, "Network-Centric Warfare: Its Origin and its Future," *US Naval Institute Proceedings* 124: 1 (January 1998) pp. 28–35 and Vice-Admiral Arthur K. Cebrowski,

"Military Responses to the Information Age," *RUSI Journal*, 145:5 (October 2000) pp. 25–29.

163. See John J. Gartska, "Defense Transformation and Network-centric Warfare," Presentation to the Australian Defence Organisation Network-centric Warfare Conference, May 20, 2003 at <http://www.defence.gov.au/strategy>.

164. For the parallel between Australian and Swedish defense expenditure see, *Sinews of War*, p. 18.

165. The most revealing sources in English are Major General C.-G. Fant, Deputy Director, Strategic Plans and Policy Directorate, Swedish Armed Forces, "Swedish Armed Forces: Towards the Future," Presentation to the ADF Military Strategy Branch, September 2002 and Lieutenant Colonel Mikael Lindberg, Joint C4I Department, Swedish Armed Forces, "Swedish Concepts for Network-Based Defense," Presentation to the Australian Defense Organization's Network-centric Warfare Conference, May 20, 2003. Both can be found at <http://defweb.cbr.defence.gov.au/strategy>.

166. Fant, "Swedish Armed Forces: Towards the Future," pp. 40–41; Lindberg, "Swedish Concepts for Network-Based Defence," pp. 16–17.

167. Lindberg, "Swedish Concepts for Network Based Defence," p. 12.

168. Ibid., p. 3.

169. Department of Defense, Senator The Hon Robert Hill, Minister for Defense, "Address to the ADF Network Centric Warfare Conference", Australian War Memorial, Canberra, Tuesday May 20, 2003, p. 1 at <http://www.minister.defence.gov.au>.

170. Department of Defense, Vice-Admiral Russ Shalders and Deputy Secretary for Strategic Policy, Shane Carmody, "Meeting the CDF's Intent: Preparing the ADF for Network-centric Warfare," Address to the ADO Network-Centric Warfare Conference, May 20, 2003 at <http://defence.gov.au/strategy>.

171. Ibid., p. 10.

172. Department of Defense, ADF Military Strategy Branch, "Enabling Multidimensional Manoeuvre: The Australian Defence Force Network-Centric Warfare Concept," Draft Discussion Paper, May 20, 2003 at <http://defence.gov.au/strategy>.

173. Ibid., p. 24.

174. Ibid., pp. 19–20.

175. Ibid., pp. 15–16.

176. Ibid., p. 14.

177. Department of Defence, *The War in Iraq: ADF Operations in the Middle East in 2003*, Department of Defence, Canberra, 2004, p. 23.

178. C. P. Snow, *Science and Government; The Godkin Lectures at Harvard University, 1960* (London: Oxford University Press, 1961) pp. 74–75.

179. See Alfred I. Kaufman, "Be Careful What You Wish For: The Dangers of Fighting with a Network-Centric Military," *Journal of Battlefield Technology*, 5:2 (July 2002) pp. 20–23.

180. Carl von Clausewitz in an 1808 essay, *Strategie* as quoted by Beatrice Heuser, *Reading Clausewitz* (London: Pimlico, 2002) p. 188.

3

All the Way with the RMA?:[1]
The Maginot Line in the
Mind of Australian Strategic
Planners[2]

Adam Cobb[3]

Introduction

Australia is roughly the size of the continental United States, yet Australia's harsh environment has set clearly defined limits on the country's development with serious strategic consequences. Australia is a vast expanse of inhospitable desert fringed by dense impenetrable jungle. The small pockets of meager agricultural opportunity that exist are fed by one small and over exploited river adjacent to the south coast.[4] Australia has an abundance of mineral resources but these are predominantly located in remote and inaccessible terrain. Access to these riches is dependent on fragile and exposed infrastructure spread across thousands of kilometers far away from any city.

Australia's unique environment has limited its population to about the same size as the population of Texas, the vast majority of whom are crowded into three big cities located in the southeast corner of the land mass. A small population base limits the resources available to the commonwealth to sustain a viable national security apparatus. The financial dilemma becomes alarmingly acute when contrasted with the sheer scale of the task of securing the massive area of the globe that Australia and its immediate strategic interests represent.

The Australian Defense Force (ADF) is charged with defending 10 percent of the Earth's surface with just 0.02 percent of the U.S. defense budget and two-thirds the number of U.S. troops based in Germany.[5] Imagine defending a resource-rich, sparsely populated country the size of the United States with less people than would fill Yankee Stadium.[6] Imagine that task in the context of an unstable local neighborhood dominated by the world's fourth most populous country, itself dominated by the biggest Muslim population on earth.

The area of immediate strategic interest to Australia extends well beyond its Exclusive Economic Zone to include as much as 30 percent of the Earth's surface, thereby appreciably complicating the task of a very small defense force. That job might be made easier if the ADF had the same kind and number of tanks, artillery pieces, aircraft, satellites, and missile systems at the disposal of U.S. forces in Germany[7] but it has drastically fewer platforms, systems, and material, much of which will be obsolete by 2015 if not sooner.[8]

These conditions represent Australia's strategic tyranny: a triple tyranny of distance, scale, and resources (human and financial), that combined, place extraordinary underlying pressures on the security apparatus of the state. Australia's security is handicapped by geography before any other consideration is taken into account.

The tyranny of distance has long given Australia strategic depth. However, this has been eroded by two key developments-one technological, one political. The gradual development and acquisition of long range precision in conventional military technologies has been gathering pace around the region. More important, however, has been the rapid emergence of global terrorism with a key hub on Australia's doorstep. This technological trend will continue to go against Australia unless the ADF's decaying Cold War legacy force structure—which was restricted by policy to limited operations in the immediate northern approaches—is modernized and enhanced to facilitate transcontinental force projection. This is as imperative for effectively moving forces around Australia as it is for sustained operations in the area. Likewise, the emergence of global terrorism will undermine any technological advances in ADF force structure.

It should come as no surprise to discover that the ADF has grasped a variety of notions of the Revolution in Military Affairs (RMA). No serious analyst would contest that the diffusion of the RMA (however defined) in Australian defense circles is well advanced and almost exclusively derived from U.S. publications and debate. This chapter touches on RMA diffusion in Australia but it is more concerned with the consequences of that diffusion. Michael Evans' chapter (chapter 2) in this volume provides a detailed description and history of departmental preoccupations with diffusing this imported concept within the defense community.

The fact that the ADF must do all it can to leverage advanced technologies against Australia's strategic tyranny—the key idea propounded by the "Knowledge Edge" policy and all subsequent defense pronouncements—is a motherhood statement and a worn one at that.[9] But that has not stopped the Department of Defense (DoD) from repeating this "RMA mantra" together with demands for new high-tech equipment in the absence of a commitment to *transform* traditional practices throughout the organization.[10] As will be argued in this chapter, genuine and thoroughgoing transformation is necessary simply to keep in the game—let alone maintain the ADF's fast eroding military superiority relative to the capabilities in the region.

The strategic tyranny, historically low levels of defense funding, the deteriorating strategic environment, a rapidly aging force structure, and excessively expensive replacement costs, all demand new thinking. The key question for the ADF is this—in a new strategic era, characterized by an emphasis on asymmetry, is it sufficient to maintain the same strategic policy, organizations, doctrine, and equipment that were applied to the unique dynamics of the Cold War? Knowing that unsatisfied powers have historically out transformed the sheltered status quo powers, the argument of this chapter is that the incremental, stultifying, bureaucratic tendencies of the DoD, have to be swept aside by innovative transformative ideas to enable the ADF to deter, defend, and defeat both old and new threats to Australia's national security.

THE ADVENT OF THE AGE OF SURPRISES

The year after the Berlin wall fell, John Mearsheimer presciently but provocatively wrote that we would soon become nostalgic for the Cold War.[11] He was referring to the certainties embodied in the bipolar standoff that kept all other international problems in check. *Uncertainty* was the key characteristic of the "post–Cold War" period. Like the interwar period of the 1920s and 1930s, the "post–Cold War" world was an interregnum, an age of anxiety, where great shifts in power were taking place that did not have a distinct predictable conclusion. A whole gamut of mostly "small" conflicts erupted all around the world. The only core theme in all these conflicts was that they took place in the periphery. Notwithstanding the decade of UN sanctioned humanitarian intervention that followed 1989, one thing was clear – the future of world history would not be determined by blue berets.

When two fully laden 767s ploughed into the twin towers of the World Trade Center on live television, the "post–Cold War" period ended. The uncertainty of the interregnum had been replaced by the ubiquitous ambiguity of terror. The age of anxiety gave way to the age of surprises. The advent of renewed global struggle was not between great powers, but rather at the intersection of the center and the periphery. The struggle is between a disparate mass of people sharing little in common beyond varying degrees of attachment to a fanatical idea dressed up as a religious duty, and the most military powerful state and alliance the world has ever seen. The tools used in the first devastating salvo of the new war were not advanced, high-tech, system of systems, satellite linked, sensor to shooter, OODA loop-crushing, long-range stand-off precision-strike, fire and forget marvels of modern military industrial might. All that was needed was some box cutters and a garden variety mode of transport.

The Global Crisis is Australia's Crisis

Australia is not immune from the shockwaves reverberating from Ground Zero. Australia is already surrounded by an arc of instability, a collection of failed states, insurgencies, economic and political basket cases, and in some cases outright anarchy. Interwoven through this volatile mix are age-old

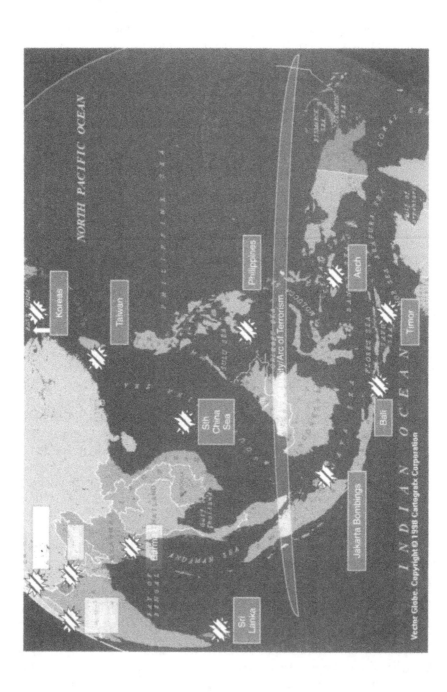

NORTH PACIFIC OCEAN

PHILIPPINE SEA

Koreas

Taiwan

Philippines

Aech

Timor

Sth China Sea

Bali

Jakarta Bombings

Sri Lanka

ANDAMAN SEA

the Arc of Terrorism

INDIAN OCEAN

Vector Globe. Copyright © 1998 Cartografx Corporation

ethnic, religious, and political enmities that were largely subdued first by Cold War competition and more recently by rapid economic growth.

Both of these overlays have been ripped away, exposing a fragile and vulnerable collection of territories that make up the arc of instability stretching from Australia's doorstep deep into the Asia Pacific region (figure 3.1). The new overlay of the global war against terrorism can only serve to further destabilize the arc and adjacent territories that are predominantly Muslim.[12]

A strategic shift against the existing order within the arc, which it could be argued was already under way as a consequence of the Asian Financial Crisis, could put very considerable pressure on Australia. By comparison to its neighbors, Australia is an isolated outpost of global capitalism, democracy, and Western values, in a sea of economic turmoil, poverty, corruption, political unrest, and growing anti-Western anger.

Sitting astride one of the world's key maritime choke points, Indonesia, is also the world's fourth most populous country and the world's largest (moderate) Muslim state. But for how long will it remain moderate? Indonesia has endured an extraordinary reversal of fortune from Tiger economy to paper tiger in just a few short years. Triggered by the Asian Financial Crisis in 1997, Indonesia's economic, political, and strategic landscape has been turned upside down. Annual per-capita income had already been reduced from US$1,200 before the crisis to $300; stock market capitalization is down from $118 to $17 billion; [and] only 22 of Indonesia's 286 publicly listed companies are considered solvent.[13]

The decline in Indonesia's GDP in 1998 was "similar to that which occurred in total during the worst years of the Great Depression (1929–32) in the United Kingdom."[14] The World Bank reported, "Indonesia is in a deep crisis. A country that achieved decades of rapid growth, stability and poverty reduction, is now near economic collapse...No country in recent history, let alone one the size of Indonesia, has ever suffered such a dramatic reversal of fortune."[15] It should not come as a surprise that the turmoil following the financial crisis led to social unrest across the archipelago. In addition to widespread riots in the capital, Jakarta, outright insurrection was unleashed in a number of provinces from Aceh to East Timor and West Papua.

The arc of instability that Indonesia straddles is replete with flash points across the threat spectrum from unstable nuclear standoff in South Asia, to a string of failed states in the South Pacific. To the long-standing disputes on the Korean peninsula, across the Taiwan straits and in the South China Sea, can be added the war on terrorism currently taking place in Afghanistan and the Philippines.[16] In addition to the existence of, or potential for, outright conflict can be added the full panoply of "new" security agenda items that so often accompany terrorist activity, such as organized crime, drug trafficking, money laundering, piracy, people smuggling, organized corruption, political intimidation, and assassination. Overlaying all of these tensions and troubles is what appears to be the contagion of the symptoms underpinning the Asian Financial Crisis. While much lip service has been paid to the recovery in Asia, many of the structural causes of regional instability remain. As the Asian Financial Crisis demonstrates, those that experience the trials and tribulations of industrialization in the late twentieth century have to accommodate rapid change in increasingly short timeframes. This naturally puts pressure on the polity and society and strains are likely to emerge, in and between states.

It's NOT About Technology, Stupid!

World War Three has become a global struggle between the opposite ends of the globalization spectrum. It is very hard to "win" a war against widespread terrorist forces that cannot be deterred or targeted like states. Advanced high-tech military forces cannot counter a widespread (mis)perception based on powerful emotions. The history of warfare has culminated in the most powerful becoming the most vulnerable. World War Three could still culminate in a nuclear holocaust, in fact it is more likely now than ever before, but just not the one we imagined 20 years ago.[17]

The significant divisions between rich and poor states also apply to the way of war among and between them. The tools of conflicts within the periphery are rudimentary and ubiquitous: famine, the AK-47, the RPG, and landmines. For wealthy states, due to a range of factors from technology to shifting international and domestic norms, the use of force has an increasingly narrow utility. In cases where poor states (and especially non-state actors) believe that they have nothing left to lose in conflicts with wealthy states, International Law and public opinion matter little to them. Indeed the development is so distinct as to suggest *a law of inverse proportion:* As the use of force becomes more limited for status quo powers, its utility grows for revolutionary powers. Similarly, the more advanced conventional military forces become, the more vulnerable status quo powers are to unconventional attack from low-tech revolutionary powers.

There are many significant limitations on the use of force in the contemporary international system for status quo powers. It is now widely accepted that the use of force cannot be legitimately resorted to unless the following conditions obtain:

- a great wrong has occurred;
- non-violent means of resolving the dispute have been exhausted;
- there is a just cause to respond;
- the UN Security Council sanctions action;
- a coalition of the willing can be formed;
- innocent casualties can be minimized; and
- an opponent's center of gravity is exposed to physical assault.[18]

These conditions rarely obtain in contemporary international disputes, with the exception of the 1991 Gulf War.

These developments limit the options available to advancd western states. Consequently, analysts have looked to technology to provide solutions in the face of tight political constraints. The trend to emphasize means over increasingly constrained ends was exacerbated by the apparent "lesson learned" from the first Gulf War. The vast expanse of flat desert that comprised the theater of operations was uniquely suited to a lightening air–land campaign—the kind of campaign the United States and its NATO allies had long prepared for and anticipated against the Soviets on the central plain of Europe.

The Gulf War was a paradigmatic war for the twenty-first century but for paradoxical reasons. Rather than being a harbinger of future wars as many RMA theorists contend, it foretells of what war will not be.[19] It demonstrated that in conventional force on force conflict (where all the above noted caveats hold) the side with the qualitative lead holds a disproportionate advantage in combat.[20] This realization had two important consequences. First, it was clear to states that could afford it that rapid military modernization was necessary to survive (deter) modern conventional combat. Second, for everyone else, it became obvious that conventional combat should be avoided at all costs. The divergence of implications of the Gulf War for rich and poor has strategic consequences.

The United States and its allies, especially Australia, became deeply engaged in efforts to understand what led the United States to its sudden unexpected victory over the world's fourth most significant military force.[21] Because the United States had such an obvious technological advantage over Iraq, that factor was quickly seized on as the fundamental key to victory.[22] That victory was so fast and so absolute over what should have been a potent adversary, many analysts came to the conclusion that technology could now be said to have a revolutionary impact on military affairs. It was little wonder then that the idea of an RMA captured the imaginations of strategic thinkers everywhere.

Some analysts argued that the revolutionary aspect of the cruise missile (for example) was to be found in the real time, integrated systems supporting its targeting, flight, navigation, mid-course redirection, real-time TV image relay of its progress, and pinpoint accuracy. Moreover, the integration of multiple layers of systems, from sensors to shooters, into a coherent whole, which could be used with considerable precision, was unprecedented in warfare.[23] On one level, of course, it did represent a quantum leap in the way of conventional war, but, on another level, it was only the latest iteration of the V-1 buzzbomb used by the Nazi's over London in the blitz.

In the case of the first Gulf War, there was an RMA but it was not what it seemed. The first Gulf War had a very significant unintended consequence. American attempts to develop a qualitative lead over the vast numerical superiority of the Soviets were *too successful.* That strategy rendered large-scale conventional warfare obsolete—at least in conflicts where one participant had a moderately well-developed "RMA force."

A Failure of Imagination

The Australian military's obsession with high-tech equipment and the dearth of asymmetric security policy development suggests that Australia has a long way to go before it is capable of transforming the Australian national security apparatus. While some elements within the Defense Department talk the language of transformation, they are preoccupied with the military means of war at the expense of policy or doctrinal innovation. The contemporary Australian DoD is mired in a stultifying culture of "learned helplessness" and groupthink.[24] The department's finances, personnel problems, force structure, acquisition strategy, and strategic policies are in various states of disarray.[25] Much stock has been placed on the RMA as the silver bullet that will solve many of the department's otherwise intractable problems.

The government allocated significant resources to research and analysis of the RMA by the DoD. A new division was established in Australian Defense Headquarters, the Office of the RMA (ORMA), headed by a one-star general, and charged with analyzing the impact the RMA would have on Australia's strategic circumstances. Initially, the organization concentrated on the impact that new advanced military technologies would have on the conventional warfighting capabilities of the current and future order of battle without any consideration for the broader organizational, military, strategic, or political implications of the RMA.

The RMA was re-badged locally as "the knowledge edge" and it immediately took pride of place in policy pronouncements.[26] The 1997 *Australia's Strategic Policy* document defined the knowledge edge as "the effective exploitation of information technologies to allow us to use our relatively small force to maximum effectiveness."[27]

It was envisaged that the knowledge edge would relieve the strategic tyranny by replacing legacy systems with new advanced military technologies that could go further, faster, carry more (and smarter) ordinance, and achieve substantially greater effects, while requiring far fewer operators, much lower life cycle costs, and fewer platform types. Unofficially some in the DoD believed that combining the knowledge edge (i.e., high technology) with the principles of effects based operations, might even facilitate a zero-sum reduction in the number of the ADFs' capabilities to fit the recent trend of reductions in the defense vote.[28] The RMA boiled down to technology enabling the ADF to do more with less.

The RMA continues to be made synonymous with advanced military technology in subsequent DoD policy pronouncements. The landmark 2000 White Paper devoted an entire chapter on the subject, tellingly entitled "science and technology."[29] However, ADF RMA-related efforts have been patchy—a few visits from important U.S. thinkers, a local conference starring American speakers, and a few papers regurgitating U.S. arguments. When an organization is so dependent on its ideas from an external source, patchy performance should not come as a surprise.

In 2003, the DoD broke with tradition[30] and publicly issued a series of concept papers and plans for experimentation that are further evidence of diffusion of RMA ideas in official circles. These short, glossy booklets are also evidence that defense is confused by the jargon associated with the U.S. RMA debate. The key documents are as follows: *Force 2020*[31]; *Future*

Warfighting Concept[32]; *Network Centric Warfare*[33]; *The Australian Approach to Warfare*[34]. The key "new" ideas raised include: Effects-Based Operations (EBOs); Multidimensional Maneuver Warfare; Network-Enabled Warfare; Network-Centric Warfare; systems thinking; and Net Assessment.

Contra their stated goal of providing a systematic exposition of RMA ideas for the ADF, from vision document (*Force 2020*) to specific planning tools (*Future Warfighting Concept*), these papers are superficial, disjointed, and lack rigor. They are unable to move beyond basic definitions and even then seem confused. For example, in *Force 2020* EBOs are defined as:

> ...the application of military and other capabilities to realise specific, desired operational and strategic outcomes in peace and war. In an Effects-Based Operation, our planning focuses on the effects that we are trying to achieve, which allows us to plan our capabilities and operations more flexibly. It avoids assuming specific platform solutions: for example, we might identify a need for "supporting fire" other than solely traditional solutions.[35]

In the *Future Warfighting Concept* EBO has a different definition:

> ...the application of military and non-military capabilities to realise specific and desired strategic and operational outcomes in peace, tension, conflict and post-conflict situations. From the military perspective, effects-based operations is more than just targeting and destroying an adversary's capacity to fight, but it also includes these aspects of warfare. It is important to understand that *effects-based operations is more about a way of thinking and planning*, and therefore about training our people, than about technology alone.[36]

Both statements suggest that Australian military planners have given no consideration to the ends ("effects") their activities have been trying to achieve. Both imply that nonmilitary means of achieving results have been ignored by strategic planners and commanders. All EBO does is *remind* planners to keep focused on the ends (the effects) they are trying to achieve and to think creatively about utilizing all means at their disposal to achieve those effects (including soft force options like psyops, cyberwar, and diplomacy).

The authors of these documents are on stronger ground when they address more traditional concerns. They admit that "Multidimensional Maneuver Warfare" is not new, but rather builds on existing maneuver doctrine. It outlines an indirect approach that emphasizes shock, tempo, agility, deception, surprise, in both the traditional and "new" dimensions (the latter being space, cyberspace, and time). "Multidimensional Maneuver Warfare" appears to be the new Australian label for "Network-Centric Warfare."[37]

In other words, advanced information technology (IT)—in particular high speed broadband data networks that share situational awareness and effective decision making among units and formations—will enable "Multidimensional Maneuver Warfare" to be realized. This is a very elaborate way of saying that the ADF plans to upgrade its IT (broadly defined), which it recognizes is a key force multiplier—an object of policy since at least the 1987 Defense White Paper.[38] There is nothing here to suggest the kind of new thinking that has been a prerequisite of past RMAs.

Only one small section out of all three currently available documents says anything new or specific. Because of its small size, and the scale and diversity of the operating environments within which it has to operate, the ADF has been able to achieve a level of jointness unparalleled in the region. Building on this substantial achievement, *Force 2020*, declares that the ADF should aspire to become a "seamless force." Becoming a Seamless Force means that:

- some units are "born joint" (tri-Service units based along functional lines);
- some force elements would be joint and inter-agency on a permanent basis;
- there will be a different and/or greater degree of joint asset management and employment. For example, Naval amphibious assets might have jointly operated helicopter support attached;
- we expect to see greater defense civilian and contractor contributions within an Area of Operations;
- we will embrace the concept of "shared stewardship and ownership." This affects all…It means that in carrying out our duties, we are not only motivated by what is best for our unit, our corps/branch/mustering, or even our Service, but what is best for our force;
- "Commonality" will be a key component: at single-Service level, "commonality" is usually associated with less of a training burden, ease of spares management, lower initial and life cycle costs, and the like.[39]

This proposal for enhancing the already strong levels of jointness within the ADF is welcome because it moves away from technological solutions and examines novel ideas about organizational change. This section is the only hint in any of these documents that the meaning of transformation as an RMA driver is understood.[40] The best that can be said of them is that these four documents represent an invitation to the organization to think about transformation. The generalized nature of the documents themselves suggest it is unlikely defense will permit any ideas that deviate too far from the norm.

Defense Policy Inhibits Transformation

Australian defense policy is paradoxical. It is deliberately designed to protect Australia against what it states is the "least likely military contingency Australia might face"—an attack on the northern coast.[41] At the same time it ignores what may already be the greatest (and most credible) threat to the country—terrorism.

Australia's defense policy is solely concerned with defending against major conventional attacks against the northern coastline.[42] Yet there is unanimous agreement that an attack on Australia is the *least likely threat* that Australia might face. Even the government's own White Paper states clearly that the key contingency Australia's defense policy is designed to counter is the one "least likely" to eventuate.[43] With respect to an invasion of Australia the 2000 White Paper acknowledges, "no country has either the intent or the ability to undertake such a massive task."[44] A major attack on Australia is judged as a "remote possibility," and minor attacks "possible," but "most unlikely."[45] No other contingencies are canvassed. Remarkably, the Defense White Paper only gives serious attention to a set of contingencies that it itself judges as unlikely and pays no attention to any other contingency.

In the only significant departure from past policy, the 2000 White Paper acknowledges that Australia "has been engaged in only one conventional conflict since the Vietnam War"[46] (the 1991 Gulf War) and goes on to make a case for a much stronger emphasis on Operations Other Than War (OOTW) following Australia's leadership in resolving the East Timor crisis and participation in countless OOTW operations in the 1990s:

> military operations other than conventional war are becoming more common. *The Government believes this is an important and lasting trend with significant implications for our Defense Force.* Over the next 10 years the ADF will continue to undertake a range of operations other than conventional war, both in our region and beyond. Many of these operations will be at the lower end of the spectrum, but often they will be more demanding. The boundary between a benign situation and open conflict can become blurred.[47]

The White Paper argues, not unreasonably, that these new OOTW demands can be undertaken within the existing force structure designed for major conventional war aimed at the Northern coast.

While the White Paper pays lips service to unconventional threats (other than OOTW) only once[48] it does not outline any strategy or make any policy prescriptions regarding unconventional war. For example, when it is noted at all, terrorism is mentioned right up there with "illegal fishing...and quarantine infringement"[49] in a passing mention of a laundry list of "new security" issues that also included "piracy...cyber attack, organised crime, illegal immigration,...[and] the drug trade."[50] *Indeed the word terrorism is mentioned only three times in the whole document.* The Defense 2000 White Paper was a classic case of preparing for the last war.

Australia's newest principal national security policy document, the *Defense 2000* White Paper (and its *2003 Update*) completely failed to address the substantial threats that have arisen against Australia and its interests in recent years. This criticism is not 20/20 hindsight in light of September 11. Australia had considerable prior warning that the nature of terrorism was changing and becoming more prevalent, from sources abroad and at home.[51] The late 1980s and 1990s witnessed a dramatic upsurge in terrorist incidents outside traditional areas of the Middle East, Northern Ireland, Columbia, and the Basque region: the Pan Am-Lockerbie tragedy, the Unabomber, the Khobar Towers attack, the Oklahoma bombing, the Tokyo Sarin nerve gas attack, the East African Embassy bombings, the first World Trade Center attack, and the attack on the USS Cole. A Presidential Commission was established in the United States to assess the threats against critical infrastructures arising from the trend in terrorism toward large-scale aggressive attacks on civilian targets. No equivalent effort was made in Australia.

In 1997, when the Defense Department was scaremongering about the Tiger economies of Asia just before they collapsed in a heap, the first ever risk assessment was conducted on Australia's critical infrastructure.[52] That assessment balanced very significant vulnerabilities in Australia's financial, telecommunication, energy distribution, and air transport networks, against the low level of threat against them. Terrorism was judged to be the most likely motivation behind any such attack. *The paper even noted that a 767 might be used in a terrorist attack against a major public building.*[53]

In subsequent reports on the same subject to the federal parliament,[54] in academic[55] and popular[56] journals, the point was raised that the DoD's single-minded focus on the RMA as a means to improve Australia's major conventional warfare capabilities was missing the point—Australia had to prepare for asymmetric contingencies because having an RMA capable defense force would limit an opponents options to unconventional warfare.

TYRANNY OF RESOURCES

Strategic planners and their political masters also made one other catastrophic miscalculation during the post–Cold War period. Believing in the rhetoric of the "New World Order," government decided to issue a peace dividend in the wake of the Cold War. Between 1989 and 1999, a total of $15 billion was taken out of the defense vote and applied to electorally popular domestic programs. The peace dividend fixed spending on domestic programs at unsustainably high levels. As domestic programs ballooned, defense personnel numbers were slashed, and forward financial programming for force structure development was postponed indefinitely. By the outbreak of the Timor crisis in 1999 Defense had spent so little on training and operations that a Parliamentary committee concluded that "preparedness has now decreased to a level where further cuts in the proportion of funding allocated to operational activities would result in unacceptable levels of strategic risk."[57]

The postponement of force structure development lead to the loss of key capabilities that should have been replaced as they reached maturity (Australia currently lacks an air warfare destroyer capability). There is now a substantial and growing set of defense capabilities that need to be replaced in a matter of a few short years and no funds for their replacement.

A double-headed crisis is undermining the ADF's long-term viability. The timely renewal of ADF capabilities is vitally important but there are currently insufficient funds to achieve this important objective. Strategic planners have turned to the RMA as a solution to this dilemma. But the kinds of advanced technologies being considered as the solution are way beyond the financial allocations currently planned for the Department. The only real solution to this problem will be to significantly increase spending and apply those funds imaginatively.

On taking office in late 1999, then Secretary of the DoD,[58] Dr. Allan Hawke, stated that the department had lost the confidence of the government and the people. Hawke also shocked his audience by admitting that the department's finances were in a "parlous" state.[59] From 1989 until 2000, the defense budget was allowed to drift downwards from 9 percent to just 7 percent of total government outlays, which equates to a trifling 1.8 percent of GDP. As the Secretary said, "The bottom line is that Australia can no longer afford a balanced, self-reliant, capable, and ready defense force of 50,000 with its current capabilities on 1.8% of GDP."[60] Defense spending is presently at its lowest level as a percentage of GDP since the Great Depression (figure 3.2).

Promises made in White Papers have seldom been converted into policy.[61] Australia has never had the budget, force structure, manpower, or the policy settings to defend itself.[62] In financial matters, defense planners have to allocate finite resources between three competing demands—operations, personnel, and capability development.

During the extended bipolar stalemate, planners focused on modernizing key maritime defense capabilities at the expense of personnel numbers and readiness. Strategic planners judged that there would be sufficient warning of a coming threat to give the ADF enough time to expand the force to meet most contingencies. But the concept of warning time was only concerned with a major military build-up aimed at Australia. OOTW and "wild card" events were not envisaged, and therefore not included in the model.

During the 1990s, the ADF had been committed to operations in the Gulf on two occasions, as well as Western Sahara, Namibia, Mozambique, Rwanda, Somalia, Croatia, Bosnia, Macedonia, Cambodia, Bougainville, Solomon Islands, and East Timor. The last four examples were very significant in terms of cost, the scale of the operation, and political risk both for Australia and the countries involved. This is an impressive list of global operations for a defense force that is specifically designed for operations in the defense of Australia.

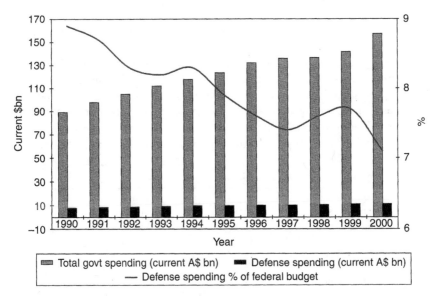

Figure 3.2 Defense spending v federal budget outlays 1990–2000.

Because these problems were unanticipated and did not fit the conventional warning model, as the operational tempo of the ADF skyrocketed, the defense budget and personnel numbers were drastically reduced. Fifteen years ago, there were more than 70,000 people in the ADF and around 40,000 on the civilian side. Today, there are some 50,000 in the ADF and around 16,000 in the department.[63] At the same time the defense budget was slashed $15 billion[64] (rounded)—which is more than the loss of a full year's funding.[65] Over the ten-year period 1989–99, federal outlays for other departments of state have escalated, Welfare spending grew a staggering 68.8 percent, followed by Health, up 52.9 percent, and Education, up 45 percent.[66]

In short, the government not only ignored its own strategic policy it did the exact opposite of what it said it would do if confronted by a series of crises. Rather than expand the force to meet the threat, the government cut the force just when it was most needed. By the time it was called on to intervene in East Timor, the ADF had become a hollow force. Had Timor not been as benign, the senior strategic planner, Hugh White, admitted to a Parliamentary committee that Australia would have "run out of" forces (and money) in the event of either combat operations or the need to sustain the 4000-strong force over a prolonged deployment (defined by White as more than nine months).[67] In other words, there was sufficient funding for the existence of the ADF, but not for its use.

When the Timor crisis emerged the major capital projects budget was raided to fund operations for the first time in history. This regrettable precedent was repeated after the outbreak of the War against Terrorism and again for Iraq. All the funds earmarked for major capability spending announced in the 2000 White Paper have disappeared into operational and personnel budgets.

BLOCK OBSOLESCENCE

The hollowness of the force might be manageable were it not for the stressful strategic environment and the incredible pressures of block obsolescence. In 1999, the department estimated that almost all major combat platforms currently in service would reach the end of their operationally useful lives by 2015. The East Timor, Iraq, Solomon Islands, and Bougainville deployments have dramatically increased wear rates on the equipment used in those theaters, bringing forward planned retirement dates.

Block obsolescence applies to the following major combat platforms:

Navy	Air Force	Army
DDGs[68]	F-111	Tanks
FFGs	F-18s	APCs
LPAs	P-3s	ASLAVs
Patrol boats	Caribou	Artillery
Older ANZAC frigates	C-130Hs	Blackhawk helicopters
Sea King helicopters	C-130Es	Chinook helicopters
Seahawk helicopters	707 AAR	Trucks and small arms

In his first speech since becoming Secretary of the DoD, Dr. Allan Hawke, admitted that by 2020 the total bill for replacing some, if not all, of these capabilities will total "between $88 and $106 billion which presently exceeds guidance for new investment by 20–40%."[69]

This is a serious calculation for replacing capabilities that will become obsolete.[70] Taking the higher figure and spreading the cost evenly over 20 years that equates to an additional $5.2 billion per year in 1999 dollars or nearly a 50 percent-increase to current outlays starting in 2000 and increasing inline with inflation over time as a crude measure of the costs involved. To date, while funding has increased it has been applied to operations, not capability development.

The political reality is that any increase in funding will be *an order of magnitude less* than the funds demanded by the department. The 2000 White Paper commits the government to spend up to an additional $500 million per year on defense following unanticipated gains in revenues from the new tax system.[71] But even if those funds were not spent on operations, as they are currently, that still falls well short of the department's demands for an *additional $4.4–5.5 billion per year for the next 20 years.*

Figure 3.3 illustrates the overcommitment of all new and existing acquisition programs. It is clear from the figure that the total major equipment program is critically oversubscribed when compared to actual appropriations.[72] The Secretary of the department has himself highlighted this problem, noting that $12.5 billion of new capital equipment (over 240 major projects compared with 160 in 1991) has been ordered between FY1996–97 and FY1999–2000. For some years acquisitions have been growing "at a rate

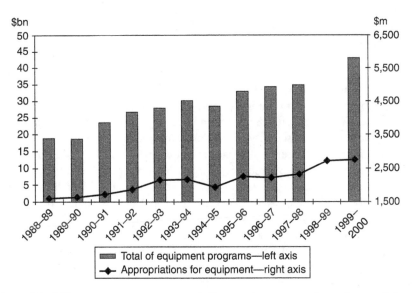

Figure 3.3 Major equipment programs—total of approved programs and annual appropriations.

significantly higher than what is affordable in the long term" according to the Secretary.[73]

As it stands today, the department is already overcommitted on new capital equipment acquisitions to a significant degree, an overcommitment that only compounds further in the light of the increasing demands of personnel and operating costs. To make matters worse, since 1989, and in particular under the current government, Australia's role in the world has grown and become more independent—witness East Timor and Australia's China policy. But glaring problems in ADF force structure, readiness, and personnel numbers are beginning to undermine the credibility of the new foreign policy.

If the department's estimates of overcoming block obsolescence are anywhere near the mark ($88–110 billion), strategic planners will need to adopt *radically* different force structures, strategic policy, and new doctrine in order to maintain a credible national security policy. Current and future governments of Australia will be looking to enhance military capabilities with fewer, smarter platforms, sensors, and munitions. Force multipliers will be critically important for this huge nation with its shrinking defense budget.

The real importance of transformation for Australia is interpreted by some as the promise of maintaining existing levels of military capability but with fewer more capable platforms. The problem with this thinking is that the advanced technologies associated with the RMA are very expensive and government appears resolute in refusing to spend any more. Superficial debates about the RMA as a solution to the department's problems is adding to the crisis.

THE TRANSFORMATION CHALLENGE

A strategic blindspot currently exists in Australian strategic policy. The stated mission of the ADF is to "defend Australia and its interests." But the strategic policy that supports this aim is focused on defending the northern coastline from conventional military attack, not defending Australia *as a whole*. To defend Australia, and not just its coastline, requires a policy and strategy for deterring and countering asymmetric and unconventional attacks against population centers, national icons, and critical infrastructures. No such policy exists. Three years after September 11, and two years after Bali, Australia has neither a national security nor a homeland security policy.

Asymmetric contingencies remain excluded from defense policy. Even the spectacular example of September 11 has not persuaded Australian planners that security does not stop at the water's edge. In fact, Australia's security policies are divided along the coastline. External security is the preserve of the DoD, internal security is shared among countless competing state and federal agencies. There are also a series of ad hoc arrangements and bureaucratic interdepartmental committees governing various elements of internal security.[74] Consequently, internal security issues do not figure in Australia's strategic policy or plans. Indeed, there is no holistic national security coordination either internally or between internal agencies and Defense.

The Chief of the Defense Force, General Peter Cosgrove, AC MC, has publicly admitted that "plainly confused communications" during the course of *Operation Relex* (to prevent the arrival of illegal people smuggling vessels in Australia) contributed to misunderstandings throughout the national security establishment (including the Government) concerning two maritime incidents. In one case, the failure of interagency cooperation led to an election scandal and a major Parliamentary enquiry. In the other case, the role of the ADF in failing to prevent the loss of hundreds of lives at sea was questioned.[75]

Expanding his comments on the failure of existing national security coordination arrangements General Cosgrove stated that

> We've got to make [interdepartmental cooperation] more robust... We've got to do better at managing the streams of communication that respond to incidents... We've got to do better with our emails. Got to somehow know where they're going and what accountability and reliability they have. Got to do better with our photographic evidence, and we've got... to ensure that we control and know about photographic evidence rather better.[76]

This is a cogent argument for the establishment of a national security council (NSC)—a permanent body staffed by senior officers—charged with synchronizing national security operational and policy coordination across government.[77] *Fusion* not separation, is the key to the transformation required to be better able to cope with the challenges of the twenty-first century.

One of the central arguments of this chapter is that Australian strategic planners have concentrated their RMA investigations on enhancing the means of war rather than examining the broader political, economic, and strategic implications of new advanced military systems. Yet even in this area of special expertise and concentrated effort, major mistakes have been made.

The textbook example of the disconnect between the ADF's seemingly insatiable desire for the very latest military technology (preferably that which is so advanced it has not yet left the drawing board) and the ability of the nation financially and technically to support the ADF's appetite for RMA systems is the COLLINS Class submarine fiasco. With Britain selling capable conventional submarines at the time, Australia decided instead to undertake a submarine building program from scratch, without any prior experience in many of the key technologies required.

The program was originally estimated to cost $5 billion but it was revealed in 2002 that costs had blown out by $3 billion.[78] To this has just been added an additional $3.5 billion over 25 years for refits and maintenance. In other words, the submarines now cost more than double the original ticket price and were acquired without any consideration of life cycle costing.[79]

The key acquisition failure was the combat system, which simply did not work. In a classic example of RMA overreach, the ADF sought a cutting-edge design incorporating a range of sensors and controls all on the one system. It was supposed to revolutionize the operation of the submarine and cut down personnel numbers. Instead, it was an abject failure. The ADF's desire to acquire the absolute cutting edge had exposed it to risks and costs it simply cannot afford and in this case left Australia without a submarine capability for some extended period of time during a period of heightened tensions.

The other classic example of ADF procurement fiascos was the LPAs that provide the ADF's amphibious capability. The East Timor crisis caught the ADF off guard—of all the strategic contingencies that planners had prepared for in the region, East Timor was one they studiously avoided. Once the crisis emerged and the ADF realized it would be deployed, it was realized that in addition to political, planning, and financial constraints, there were deep gaps in ADF heavy sea and air-lift capabilities.[80] These gaps were filled in the short term by leasing a fast catamaran from InCat ship builders in Tasmania, and Russian Antonov aircraft and crews.

The catamaran HMAS JERVIS BAY could move 500 fully equipped troops with their vehicles, including armored personnel carriers, light armored vehicles, and trucks. The boat's maximum range is approximately 1,500 nautical miles, at speeds of more than 40 knots. JERVIS BAY'S commanding officer, Lt. Cdr. Jonathan Dudley, RAN., stated during the Timor Operation "we can make up to three runs a week between here (Darwin) and Dili, East Timor," Dudley added that JERVIS BAY'S crews have made the 430 nautical-mile (~1,000 km) route between Darwin and Dili a total of 74 times in the 12 months from September 1999 when the operation commenced. "It's really quite amazing, especially when you consider our capacity on each trip."[81] After Timor, however, the ADF abandoned the catamaran.

The value of such vessels was not lost on the U.S. Navy (USN). The USN is currently trailing several new InCat vessels in the Persian Gulf and sees it as a key transformational capability for the USN in the twenty-first century. Moreover, InCat is working on a range of new military applications of these vessels from carriers to littoral combat ships. Compared to the existing technology' these ships are fast, economical, much less personnel intensive, and offer the possibility of stealth design.

In stark contrast, pre-Timor, the RAN had rushed to purchase two mothballed and outdated U.S. amphibious vessels offered by the USN on the cheap. The Navy paid just $61 million for both ships, slightly less than the cost of one InCat ship. Because the condition of the old USN ships was significantly worse than expected, the total cost of the project was calculated by the ANAO to escalate to $445 million.[82] An additional $60 million should be added to that as the cost of leasing the Timor catamaran while the ex-USN ships were in dry dock. On top of that, the USN ships require a crew of 180 whereas the catamarans only require 25.

So for half a *billion* dollars the RAN acquired two outdated, slow moving, old technology ships, due to be retired in ten years after their purchase, while the USN is investigating Australian state-of-the-art ship-building technologies at around $100 million a copy. RMA diffusion has been essentially a one way street between the US and Australia. Ideas, technologies, source codes (when released), CONOPS, and jargon, have all flowed in one direction. The "Fast Cats" are an important and significant contribution to generating a flow in the other direction in the terms of new technology requiring new doctrine. However it is lamentable that elements within the ADF refuse to champion a key indigenous contribution to the next evolution in naval concepts and operations. This clearly demonstrates that even the technological aspects of the RMA are not well understood within the ADF.

CONCLUSION

Advanced rich states no longer need to resort to the use of force to resolve conflicts. Nor can they afford modern conflict in the face of conventional mutually assured destruction wrought by the advent of the RMA. The poor will continue to kill one another far away from the prying eyes of the world. The future of major conventional and terrorist conflict will be between rich and poor. In such conflicts, the center of gravity will rarely be physical for either side and thus the utility of conventional force will decline in favor of alternative strategies. This does not bode well for a deep reliance on the technological superiority of conventional military forces at the heart of the RMA concept. Deep transformation of not just defense forces, but the whole way of perceiving and responding to security issues, must occur.

Australian defense policy has left the country in a double jeopardy— Australia is open to *both* asymmetric attack (due to policy blindspots) and in the near future, conventional attack (as extant capabilities decline and tensions mount). Focusing *only* on developing an RMA force exposes a weakness in defense against unconventional security threats. Yet, at the same time, mistakes, delays, and misallocation of vitally needed funds in the hot pursuit of a poorly conceived RMA force structure can potentially undermine deterrence and invite a conventional attack.

The Australian obsession with a purely technological interpretation of the RMA ignores the much bigger issue of defense transformation. ADF planners must reintroduce Clausewitz into their thinking. Instead of constructing modern day Maginot lines of advanced military technology as an ultimate and impregnable shield designed for just one task—the fundamental questions of who, what, where, how, and why, have to be re-considered. Not only must the deeper political questions be asked but consideration of methodologies, management structures, means, and ends must be comprehensively addressed.

Most opponents will seek asymmetric advantages by side-stepping a comprehensive and well-structured RMA force if they are able. That does not mean, however, that conventional forces can be forgotten and left to decline. If a conventional force has significant weaknesses, more often than not, the path of least resistance leads to effective and spectacular attacks against the force as in Somalia, Vietnam, and Afghanistan.

Paradigm Shift?

Australia suffers adversity, scarcity, and crisis in abundance. Some elements within the DoD talk about transformation but there are serious questions whether the meaning of transformation is properly understood and whether the organization would accept the kind of radical ideas that have proven necessary in the past to stimulate an RMA.

Clearly much work still needs to be done if the DoD is ever to attempt an RMA prior to conflict. That leads to a final paradox: it is necessary to have an advanced and comprehensive RMA force in order to reduce the number and effectiveness of options available to an opponent. Having achieved a state of comprehensive conventional deterrence, it is then necessary to devise innovative strategies to deny the opponent an asymmetric advantage. Agility of the mind and not a high-tech system of systems will be the fundamental first priority of the sophisticated strategist of the early twenty-first century.

NOTES

1. At a joint press conference between the two leaders during the U.S. President's tour of Australia in 1966, Prime Minister Harold Holt pledged that Australia was "all the way with LBJ" in the war in Vietnam. The phrase has since become iconic in Australia for our attitude toward the alliance with the United States.
2. My thanks to Australia's leading air power strategist, Dr. Carlo Kopp, Visiting Research Fellow in Air Power and Military Strategy at ADSC (UNSW at ADFA), for his comments on earlier drafts of this work. I am also very grateful to Professor Tom Mahnken of the U.S. Naval War College and Professor Emily Goldman of UC Davis for their valued feedback, encouragement, and patience during the preparation of this chapter. All mistakes are my own.
3. Dr. Adam Cobb is a former senior defense adviser to the Australian Parliament, and Special Director—Strategic Policy at AFHQ. Cobb is founder and director of Stratwise Strategic Intelligence, Australia's first privately funded strategic think tank and advisory service (see www.stratwise.com).
4. Arable land comprises around 6% of the land mass.
5. According to the U.S. DoD website in FY2001 the total U.S. Defense budget was U.S.$291.1 billion. The Australian Defense budget for the same year was Ca. U.S.$6 billion. For the United States see: http://www.defenselink.mil/ news/ Feb2000/b02072000_bt045-00.html, for Australia see: http://www.Defence. gov.au/budget/. ADF personnel stands at around 50,000, U.S. deployed troops in Germany stand at 71,258, see http://web1.whs.osd.mil/mmid/m05/ hst0601.pdf.
6. According to http://www.ballparks.com/baseball/american/yankee.htm Yankee Stadium's capacity is currently 57,545, which is ~4,000 more seats than the ADF has people in uniform!
7. Or adjacent to Germany, in the case of ships.
8. Indeed, the requirement to modernize the ADFs force structure may arise a lot sooner than Defense's 2015 estimate because of the "unforeseen" leap in optempo that has occurred since that estimate was made.
9. References to the power of technology in overcoming elements of Australia's strategic tyranny date back to Federation in 1901 and in some detail since the seminal 1976 Defense White Paper.
10. The series of concept papers issued by Defense in 2003, listed later, are a very significant advance on the White Paper (2000 and 2003 update) yet they disappoint by failing to move beyond basic definitions and generalizations. Had they

made a set of explicit funded proposals then their value would have been greatly increased but they had no traction within the department and the Defense Capability Plan put before Cabinet in late 2003 utterly ignores the RMA and defaults back to demanding more of the same—i.e., Cold War legacy systems. See, for example: *Future Warfighting Concept; The Australian Approach to Warfare*; and *Force 2020*, full references later.

11. J. Mearsheimer, "Why We Will Soon Miss The Cold War," *The Atlantic*, vol. no. 266 (August, 1990) pp. 35–51.
12. For a comprehensive catalogue of Osama bin Laden's threats and plans see, R. Gunaratna, *Inside Al Qaeda: Global Network of Terror* (Melbourne: Scribe, 2002).
13. P. Dibb, D.D. Hale, and P. Prince, "'The Strategic Implications of Asia's Economic Crisis.' Survival," 40: 2 (1998) p. 11.
14. H. Hill, "An Overview of the Issues," in H.W. Arndt and H. Hill, ed., *Southeast Asia's Economic Crisis: Origins, Lessons, and the Way Forward* (Sydney: Allen and Unwin, 1999) pp. 1–8 (emphasis in the original).
15. World Bank, *Indonesia in Crisis: A Macroeconomic Update* (Washington, D.C. World Bank, 1998) p.1 (emphasis added).
16. Of course, the struggle against the Abu Sayyaf in the Philippines has been going on for decades but it has recently taken on a new dimension in light of the attacks on the United States.
17. The incredible history of nuclear mistakes not withstanding.
18. This criterion is particularly important and often overlooked.
19. At least between rich and poor. War between the poor will continue its futile course, and war between the rich has been rendered obsolete militarily by the RMA and strategically by a long and complex range of political and economic developments.
20. This statement holds so long as the set of conditions noted at the start of this section holds. terrain is amenable to conventional warfare and there is not some kind of disproportionate support for the underdog (as in Vietnam). The qualitative lead could be across a range of factors, the will to win, training, strategy, technology, force structure, command and control, and so on.
21. In addition to the sheer size of its army, Iraq possessed quite capable military technologies of its own including first-tier Soviet military hardware such as AEW&C aircraft, SAM systems, and electronic warfare capabilities.
22. It also helped that the impact of technology was easier to measure than the other qualitative advantages the U.S. forces enjoyed over their opponents and dovetailed neatly with the technologically driven interests of military planners.
23. William Owens (Former Vice Chairman JCS) *Lifting the Fog of War* (New York: Farrar Straus & Giroux, 2000).
24. "Learned helplessness" was used by the Secretary of Defense to refer to the breakdown of professional responsibility within the DoD. See Dr. Allan Hawke, Secretary of Defense, "What's the matter—a due diligence report," Address to the Defense Watch Seminar, February 17, 2000. I. Janis, *Groupthink*, 2nd Ed., (Boston: Houghton-Mifflin, 1982). Lieutenant Colonel Neil James, provides a very thorough and historically insightful explanation as to why groupthink and learned helplessness have such a strong grasp on the DoD mind. His report is highly contentious because it challenges the orthodoxy that inhibits original thought within the Defense organization. See James, op. cit.
25. As evinced in countless Parliamentary and Australian National Audit Office reports on various deficiencies in the management of the DoD.

26. "The Knowledge Edge" was the key new idea introduced by the 1997 mini-White paper entitled *Australia's Strategic Policy*. Issued just months before the Asian Financial Crisis, this document argued that the Asian economies were growing so large and powerful that it was only matter of time before their greatly enlarged internal revenues would be applied to Defense acquisitions, which in turn would enable them to threaten Australia and its interests.

27. *Australia's Strategic Policy*, p. 56.

28. This observation was made by the author when he was working in the DoD.

29. *Defence 2000* (Canberra: DoD, AGPS, 2000) chapter 10, pp. 107–113.

30. Australia does not have a tradition of publishing official concept documents. Australia's military "think tanks," such as the Sea Power Center (http://www.navy.gov.au/spc/), the Land Warfare Studies Center (http://www.Defence.gov.au/army/lwsc/), the Aerospace Centre (http://www.Defence.gov.au/RAAF/aerospace/), the Australian Defence Studies Centre (http://idun.its.adfa.edu.au/ADSC/), and the Australian Strategic Policy Institute (http://www.aspi.org.au/), publish doctrinal and other studies, yet they rarely gain the same attention as official announcements.

31. Department of Defense, *Force 2020*, Defence Public Affairs and Corporate Communication, June 2002. (http://www.Defence.gov.au/publications/f2020.pdf).

32. Department of Defense, (Policy Guidance and Analysis Division), *Future Warfighting Concept*, Defense Public Affairs and Corporate Communication, December 2002. (http://www.Defence.gov.au/publications/fwc.pdf).

33. Department of Defense, *Network Centric Warfare*, Defense Public Affairs and Corporate Communication, *Forthcoming*.

34. Department of Defense, *The Australian Approach to Warfare*, Defense Public Affairs and Corporate Communication, June 2002. (http://www.Defence.gov.au/publications/taatw.pdf), p. 24.

35. *Force 2020*, p. 22.

36. *Future Warfighting Concept*, p. 12 (emphasis added).

37. Much like the RMA was relabeled the Knowledge Edge—the important point is further indigenous development of the concept fails to occur beyond relabeling.

38. *The Defence of Australia*, a Policy Information Paper presented to Parliament by the Hon. K.C. Beazley (Canberra: Australian Government Publishing Service, 1987), p. 69.

39. *Force 2020*, p.18. The test for the force-wide commonality goal is already upon the department in Project Air9000 that aims, inter alia, to rationalize the ADF helicopter fleet, which currently operates eight different helicopter types for a Defense force of 50,000.

40. Moreover the author is skeptical about the notion of testing highly subjective propositions in the first place. The structure and methodology of the wargames attended by the author suggest that just about anything can be derived from an ADF wargame depending on the predispositions of the participants, the factors included or excluded and the subjective means of deriving a conclusion to the game. Defense seems to equate a wargame with the testing rigor of a scientific experiment when the facts simply do not support the case.

41. *Defence 2000* (Canberra: DoD, AGPS, 2000) p. 23.

42. Geography suggests the Northern orientation.

43. *Defence 2000*, (Canberra: DoD, AGPS, 2000) p. 23.

44. Ibid.

45. Ibid.

46. Ibid, p. 10.

47. Ibid, p. viii (emphasis added).

48. This comment is made in the introduction and repeated once more in the body of the text.

49. Ibid.

50. Ibid.

51. Since the end of the Cold War, terrorism was becoming more violent, its political basis shifting from making statements to inflicting maximum damage, it was becoming more indiscriminate and was using new tools such as Sarin gas.

52. A.C. Cobb, "Australia's Vulnerability to Information Attack: Towards a National Information Policy," *Strategic and Defence Studies Centre, ANU, Working Paper*, No. 306 (1997).

53. Ibid., p. 24.

54. A.C. Cobb, *Thinking the Unthinkable: Australian Vulnerabilities to High Tech Risks*, Parliamentary Report #18 (1998).

55. A.C. Cobb, "Electronic Gallipoli?," *Australian Journal of International Affairs*, 53:2 (1999).

56. A.C. Cobb, "Electronic Gallipoli?" *Quadrant* (April, 1999).

57. Joint Standing Committee on Foreign Affairs, Defence and Trade, *Funding Australia's Defence* (April 1998) p. 85.

58. In the Australian system the Secretary of the department holds equal power with the Chief of the Defense Force. While their job titles might suggest different functions officially they are both charged with leadership of the department and forces as a whole. Moreover there are two separate Headquarter systems—one for Administration (HQADF) and one for Combat (HQ Australian Theater).

59. Dr. Allan Hawke, Secretary of Defense, "What's the Matter—A Due Diligence Report," Address to the Defense Watch Seminar, February 17, 2000; Dr. Allan Hawke, Secretary of Defense, "Money Matters," *Address to the Royal United Services Institute of Victoria* (April 27, 2000) p. 12; Dr. Allan Hawke, Secretary of Defense, "One Year On," *Address to the Defense Watch Seminar at the National Press Club*, February 27, 2001.

60. A. Hawke, Money Matters, (http://www.defence.gov.au/media/speechtpl. cfm?CurrentId=921)

61. For example, both the 1987 and 1994 Defense White Paper's recommended funding increases to bring defense spending to the level of 2.6–3% and 2% of GDP, respectively. Yet in 1987–88 the actual outlay was in fact *cut* by 1.1%. Joint Standing Committee on Foreign Affairs, Defense and Trade, *Funding Australia's Defence* (Canberra: AGPS, 1998) p. 20; 27.

62. There was a period in the 1980s when "defence self reliance" was talked about but only within the context of "an alliance."

63. Hawke, "What's the matter—a due diligence report," op. cit.

64. See *Funding Australia's Defence*, op. cit, p. 4.

65. The $15 billion was arrived at using the following technique: establish how much 9% of the federal budget for was each year and subtract the actual amount spent on defense—then add the sum of the remainder over the ten-year period.

66. See Joint Standing Committee on Foreign Affairs, Defense and Trade, *Funding Australia's Defence* (Canberra: AGPS, 1998) p. 4.

67. Mr. Hugh White, then Deputy Secretary Strategy, comments during a "Defence Strategy Debate" held by the Defense subcommittee of the Joint Standing

Committee on Foreign Affairs, Defense and Trade, Parliament House, Canberra, June 30, 2000. Hugh White is now the Director of the new Australian Strategic Policy Institute, chaired by Professor Bob O'Neill.

68. The DDGs have already all been decommissioned leaving Australia without an air warfare destroyer capability.

69. "Defence—the State of the Nation" Speech by Dr. Allan Hawke, Secretary of the DoD to the United Services Institute, February 2, 2000, p. 10.

70. This was confirmed by senior defense financial officials who wished to remain anonymous. Interviewed February 17, 2000.

71. In 2000, the federal government introduced a 10% broad-based "Goods and Services Tax," which has delivered a revenue windfall substantially above government predictions.

72. Figure kindly supplied by Derek Woolner.

73. Hawke, "What's the matter," op. cit.

74. Such as the Standing Advisory Committee on Commonwealth/State Cooperation for Protection Against Violence (SAC-PAV).

75. Two examples have recently come to receive widespread public attention. Both involve the recent spate of illegal people smuggling vessels attempting to land in Australia. In one case, refugees were accused of threatening to throw their children overboard to gain attention from the Navy, causing a political furore when partial information about the incident was announced to the press in an election campaign (SIEV-4 and the "Children Overboard Affair"). In the other case, a similar vessel sank with the loss of 353 lives. In addition to confusion as to whether the ADF was surveilling the craft and therefore may have been in a position to mount a rescue, the exact location of the vessel when it sank is in dispute (the "SIEV-X incident"). SIEV stands for "Suspected Illegal Entry Vessel."

76. Address to the National Press Club, July 30, 2002.

77. The creation of an Australian NSC was not the new Defense Chief's objective but it flows from the problems he identified and the solutions he suggested.

78. ABC TV "7.30 Report" 3/9/02, http://www.abc.net.au/7.30/s666363.htm

79. Minister For Defense Press Release, "ADELAIDE WINS MULTI BILLION DOLLAR SUBMARINE REFIT CONTRACT" Thursday, October 16, 2003 131/2003.

80. A.C. Cobb, "East Timor and Australia's Security Role: Issues and Scenarios, Current Issues," Brief 3, Parliamentary Research Service 1999.

81. September 2000, USN news report. See http://www.c7f.navy.mil/news/2000/09/16.html

82. Australian National Audit Office (ANAO), *Amphibious Transport Ship Project*, ANAO Report 8, 2000–01, p. 12.

4

THE JAPANESE PERCEPTION OF THE INFORMATION TECHNOLOGY-REVOLUTION IN MILITARY AFFAIRS: TOWARD A DEFENSIVE INFORMATION-BASED TRANSFORMATION

Sugio Takahashi

INTRODUCTION

In the early 1980s, a segment of the Soviet armed forces led by Marshal Nikolai Ogarkov predicted that the rapid development of computer systems and high-technology weapons by NATO would lead to drastic changes in how warfare was waged.[1] Ogarkov referred to this transformation as a "Military Technological Revolution."[2] Strategists around the world agreed that a revolutionary change in warfare was underway[3] and that a key aspect of this "Revolution in Military Affairs" (RMA) was the information revolution.

This prophecy has been fulfilled. The initiator of revolutionary change, the United States, has conducted four major military operations since 1990: *Desert Storm* in 1991, *Allied Force* in 1999, *Enduring Freedom* in 2001, and *Iraqi Freedom* in 2003. The United States secured decisive military victories in each of these operations, demonstrating its dominant military strength to the world.

The debate surrounding the RMA has become a very important part of strategic studies even in Japan. Research was conducted within several strategic institutes, and the Japanese Defense Agency (JDA) carried out its own study of the RMA. As its study shows, the JDA is seeking a version of the Information Technology-RMA (IT-RMA) relevant to Japan's specific situation. Since a nation's force structure and defense posture are functions of its political/strategic situation, Japan needs a roadmap for its IT-RMA that differs from the one that the United States is following. This chapter compares the strategic situations facing Japan and the United States, and demonstrates why Japan requires its own unique version of the IT-RMA.

A CONCEPTUAL ANALYSIS OF THE RMA

Many strategic experts have identified a revolution within the U.S. military, although they use different terms: military revolution, RMA, and defense transformation. It is important to clarify the meanings of these concepts.[4]

The concept of military revolution emerged from a debate among military historians[5] about whether changes in European warfare in the sixteenth and seventeenth centuries produced changes in European society. Michael Roberts argued that the emergence of a standing army and the decline of mercenary armies led to the development of taxation and bureaucratic systems to raise and support a mass army. Proponents of this argument understand "military revolution" as a transformation in society and international relations led by changes in military affairs. In the present-day context, "military revolution" implies that ongoing information-based military modernization will spill over and impact society and international relations.

The term RMA is narrower, literally implying a change *in* military affairs while excluding changes *outside of* military affairs. Some RMAs, such as the Napoleonic Revolution, can be thought of as military revolutions but an RMA does not necessarily imply a change in society or in international relations. The *Dreadnought* revolution, aircraft carrier revolution, and *Blitzkrieg* each radically transformed the battlefield but did not spill over into areas outside of military affairs.

A third concept, defense transformation, emerged in the late 1990s and was first used in the report, *Transforming Defense: National Security in the 21st Century*, issued by the U.S. National Defense Panel in 1997.[6] Since the inauguration of George W. Bush, defense transformation has guided U.S. military modernization efforts.[7] Defense transformation is the process for reaching the RMA (or military revolution). Whether the ongoing defense transformation is revolutionary or not will be determined by future military historians. In this chapter, defense transformation refers to process and RMA refers to goals.

Three Innovations in the Current RMA

The ongoing U.S. defense transformation consists of innovations in three areas: weapon systems, information systems, and doctrine/organization.[8] The main weapon systems innovations are precision guided munitions (PGMs) and stealth aircraft. Information system innovations include the development of advanced sensors and information networks to transmit data from these sensors—AWACS, JSTARS, satellite systems, and the Joint Tactical Information Distribution System (JTIDS). Organizational and doctrinal innovations involve transformations of relations among actors in a military organization and the development of tactics. U.S. organizational/doctrinal innovations include the bid for jointness after the Goldwater–Nichols Act of 1986, and development of a joint operational concept in *Joint Vision 2010* of 1996 and *Joint Vision 2020* of 1999.

These three areas of innovation began over a decade ago. Development of the Tomahawk cruise missile and JTIDS started in 1972 and 1974, respectively. The acquisition of F-117A and the development of JSTARS began in 1982. Without these consistent efforts over some decades, the United States would not have achieved its current military advantage, although these programs were developed in an effort to win the Cold War, not regional conflicts in the post–Cold War era.

One country's innovations may not make sense for other countries operating in different strategic environments. For the United States, it is quite reasonable to improve its rapid-deployment capability to cope with regional conflicts, construct a joint information network to connect dispersed units, and pursue standoff PGM-intensive warfare by utilizing its advanced information and defense technology. But the U.S. style of transformation, such as a rapid-deployment capability, does not make sense for other countries that do not need to project power. Or if a country does not have access to the technology for an IT-RMA, transformation similar to that underway in the United States is an unattainable goal. For other countries, the U.S. path to the IT-RMA is not a generalized roadmap for achieving national security.

The External Effect of the "American" RMA

Military organizations constantly seek to strengthen their capability in accordance with their strategic perception, technological capabilities, and available resources. These differ from country to country, so the optimal defense posture will by necessity be different from country to country. Consider the military innovations that occurred during the period between the two world wars.[9] Among the three major naval powers during the interwar period, Britain did not develop equipment and tactics for effective offensive use of the aircraft carrier while Japan and the United States built tremendously capable carrier battle groups.[10] The Royal Navy at that time faced limited resources and lacked immediate threats that required highly capable naval aviation. In this way, the Strategic Situation determines the character of an individual country's defense posture. In addition, even if countries require revolutionary change, they will not necessarily copy other countries. Again, the case of the aircraft carrier is instructive. Only two countries (Japan from 1941 to 1944 and the United States from 1942 to the present) have operated carrier battle groups effectively in offensive operations. Certainly aircraft carriers brought about a huge change in the force structure of the U.S. Navy. But even six decades after Pearl Harbor, the fleets of many countries are still comprised of surface ships, not aircraft carriers, because carrier battle groups are tremendously expensive. In the current context, most countries do not face the same strategic situation as the United States, do not possess its level of advanced technology, and do not have comparable resources. These countries will not likely follow the U.S. path in defense transformation.

In the present strategic situation, the first priority for the United States is dealing with regional conflicts and "rogue states." Rogue states may possess weapons of mass destruction (WMD), but their conventional forces are far less lethal than were the Soviet Union's. The heavy-equipment-based forces that the United States fielded to fight against the Soviet Union have become relatively useless because they are difficult to deploy quickly to deal with unexpected regional conflicts. Though the "two Major Regional Contingency" or "two Major Theater War" force structure, stipulated in the 1993 Bottom-Up Review and 1997 Quadrennial Defense Review, was intended to prepare the United States for regional conflicts in the Middle East and on the Korean Peninsula, the next two major military operations after the Bottom-Up Review were conducted in the Balkans and Afghanistan. While heavy but less deployable forces are less useful today, light and rapidly deployable forces lack the lethality of heavy forces. To meet the contradictory requirement for lethality and rapid deployment, the United States has promoted defense transformation. But not all countries face the same strategic requirements as the United States. Japan is a case in point. Japan has highly sophisticated information technology, but its strategic requirements are quite different from those of the United States. A rapid-deployment capability is not necessary for Japan's Self-Defense Forces (SDF) because Japan does not dispatch troops for combat beyond East Asia. Its defense policy is based on the "Exclusively Defense Oriented Policy."

Potential U.S. adversaries also have different defense requirements—to counter U.S. power projection—and are likely to emphasize an "access denial capability." The most effective solution for many countries is to avoid an information-based transformation, which requires sophisticated IT and huge amounts of resources, and to acquire nuclear weapons to deter the United States. Future U.S. adversaries may pursue a "nuclear RMA" (perhaps resembling Eisenhower's "New Look Strategy"). Ironically, the American quest for an IT-RMA may result in further proliferation of WMD, especially nuclear weapons. The lethality of precision forces in absolute terms can never match that of nuclear weapons and nuclear force will remain the ultimate deterrent.[11] This, in turn, implies an increased need for missile defense systems to protect the United States and its allies.

The probability that other countries will seek to replicate the defense transformation of the United States is not high. Rather, potential adversaries may pursue an "asymmetrical" response. The United States and other followers of the IT-RMA must take these "external effects" of American defense transformation into account. The next sections focuses on Japan's strategic situation and its unique path for defense transformation.

JAPAN'S DEFENSE POLICY

Each country builds its military forces to cope with its own strategic agenda given available resources. Even within an alliance, a member country may pursue a different transformation path than the rest of the alliance. The IT-RMA may provide a splendid military capability in absolute terms, but it is useless if that capability is not relevant to the country's strategic situation.

Japan's Defense Policy Framework

The ongoing debate on the IT-RMA implicitly assumes the U.S. strategic condition. But other countries, like Japan, have different strategic situations and so may pursue different forms of the IT-RMA. Japan's defense policy, represented by the "exclusively defense-oriented policy (EDOP)," is completely different from the American defense policy.[12] This policy was formed in the 1950s under the Constitution and stipulates that Japan's SDF are organized for passive defense only. Based on EDOP, the SDF's military plans and defense strategy have always focused on territorial defense. Also, the Diet prohibits the SDF from acquiring "exclusively" offensive capabilities, such as aircraft carriers, long-range strategic bombers, and ballistic missiles.

Not only is Japan prohibited from possessing "exclusively offensive" capabilities under the EDOP, but Japanese leaders have voluntarily restrained themselves from acquiring certain capabilities. A land attack capability provides a good example. The Japan Air Self-Defense Force (JASDF) has limited land attack capability and has concentrated its resources on interceptors. Currently it has about 200 F-15s as interceptors and about 50 fighter-attackers, both indigenously developed F-1s and F-2s jointly developed with the United States. The uniqueness of these fighter-attacker aircraft is that land attack—including close air support—is not their primary mission. This fighter-attack force was developed as an anti-ship force to interdict an invader's amphibious forces. This unique resource allocation strategy follows directly from the EDOP. Also, the JASDF does not possess land-attack PGMs. Nor has it traditionally had aerial refueling planes, although the Diet admitted to acquiring them in 2001.[13] A few radical opinions are calling for Japan to acquire offensive capabilities, like the Tomahawk cruise missile, in response to North Korea's nuclear weapons and ballistic missiles, but there is as yet no official movement in this direction.

The SDF also has a limited power projection capability. The air wing of the Japan Maritime Self-Defense Force (JMSDF) consists of helicopters and P-3Cs and focuses exclusively on anti-submarine warfare (ASW). Japan cannot achieve air-superiority beyond the range of JASDF fighters. Since Japan is composed of numerous islands, the JMSDF has landing and helicopter-capable vessels for maritime transport, but there are no marines in the SDF, and the SDF has no amphibious capability that can be used to invade other countries.

The Japan–U.S. alliance has also been crucial in shaping Japan's defense policy. The uniquely passive nature of the EDOP implicitly assumes U.S. support. Under the current framework of the Japan–U.S. alliance, Japan provides bases for U.S. Forces in Japan (USFJ) and, in a regional contingency, would conduct logistics support and defensive operations to protect Japan's territory and USFJ. The United States is responsible for extended nuclear deterrence and offensive operations. This division of labor is clearly stipulated in the 1997 "Guidelines for Defense Cooperation between Japan and the U.S," the basic document for bilateral operational cooperation.[14]

In sum, Japan acts as the "shield" and the United States acts as the "spear." This division of labor differs completely from the NATO and U.S.–South Korea alliance, in which military units operate jointly under unified command. The SDF and the USFJ coordinate operations but operate separately without unified command.

The EDOP and the Japan–U.S. alliance will continue to be the primary pillars of Japan's defense policy, and will be decisive to Japan's vision of the RMA. Maintaining a passive defense posture and interoperability with the United States are the key principles guiding the Japanese version of the RMA.

Other Factors Influencing Japan's IT-RMA

In addition to Japan's defense policy framework, several other factors will influence Tokyo's approach to the RMA: technological advantage, the regional military balance, the expanding role of the SDF, social sensitivity to casualties, demographic trends, and the financial situation.

Japan is one of the leading technological nations in the world and has the wherewithal to pursue the RMA, but there are problems in its defense industry. Japanese defense corporations do not occupy a significant position in the global defense industry because of the country's severely constrained defense export policy that does not permit its defense industry to export products. Japan was not a player in the transnational defense industry that unfolded after the Cold War.[15] While Mitsubishi Heavy Industry is the largest defense corporation in Japan, but ranked 14th in the world defense industry in 2002 based on defense-related revenue. The second largest firm in Japan, Kawasaki Heavy Industry, ranked 39th and the third largest, Mitsubishi Electric, ranked 52nd.[16]

Another unique aspect of Japanese defense industry is that defense-related revenue occupies a very limited share of the total revenue of Japan's defense corporations. Mitsubishi Heavy Industry's defense-related revenue was 13.4 percent of total revenue in 2002. Kawasaki Heavy Industry's defense-related revenue was 8.9 percent and Mitsubishi Electric's was 2.0 percent. On the other hand, Lockheed Martin, the world's largest defense contractor, had a defense-related revenue of 87.8 percent of total revenue. Raytheon, the third largest defense contractor, had defense-related revenue that was 91.2 percent of total revenues. Even for the prominent manufacturer of commercial airplanes and the second largest defense contractor in the world, Boeing, the share of defense-related revenue reached 40.8 percent. Mitsubishi Heavy Industry's total revenue is larger than Raytheon's total revenue, and Mitsubishi Electric's total revenue is larger than Lockheed Martin's total revenue. So Japan's defense corporations may not rank highly in the defense industry, but they have performed consistently well. This minor share of defense-related revenue has an important implication for the IT-RMA, because of the important role played by "dual-use" technologies. Since Japanese defense industries concentrate on civilian technology, they can utilize civilian "dual-use" technology in pursuit of the IT-RMA.

The regional military balance will also shape Japan's approach to the RMA. Since military forces exist primarily to prepare for external threats, Japan's military planners must take into account its neighbors' military buildups. For the time being, only the United States, Western Europe, and Japan have the greatest technological potential to pursue the IT-RMA, so in the short term, Japan needs to prepare against pre-RMA threats in the region. Countries that do not possess IT-RMA potential will likely seek to acquire WMD. Consequently, countermeasures against nuclear proliferation, such as ballistic missile defenses (BMD), merit high priority in Japan's RMA.

In addition to warfighting, the roles and missions of military forces have expanded since the end of the Cold War to include "military operations other than war" (MOOTW). The SDF has participated in domestic disaster relief operations, and in peacekeeping operations in Cambodia, Mozambique, the Golan Heights, and East Timor. The SDF has provided logistical support for the global war on terrorism in the Indian Ocean and works for post-conflict reconstruction in Iieq. This expansion of roles and missions will continue so Japan's plans for the RMA must consider the requirements for international MOOTW such as strategic transportation and interoperability.

A variety of societal factors will also influence Japan's approach to the RMA, particularly social sensitivity, demographic trends, and the financial situation. Japan's society is highly casualty-sensitive, perhaps more than other advanced countries, because of anti-military sentiments after the Second World War and a high regard for human life.[17] Limiting the vulnerability of SDF troops and preventing civilian collateral damage must be accommodated in Japan's RMA vision.

Demographic trends merit serious consideration. The birthrate in Japan has been declining. The SDF must be able to address retention problems in the near future. To avoid a decline in capability, the SDF must improve the operational performance of individual units and the IT-RMA may be a solution to this problem. Introducing information technology allows the SDF to maintain its operational capability with a smaller pool of recruits.

Finally, visions for the IT-RMA may crumble like sand due to Japan's fiscal condition. In the FY2003 budget, the revenue from national bonds occupied 44.6 percent of revenue, and the accumulated sum of the national bonds reached $3.75 trillion, which is about 70 percent of Japan's annual GDP. So, the flexibility of Japan's budget has been limited. It is especially difficult to increase a defense budget that occupies about 10.4 percent of the national budget ($41.2 billion).[18] Cost-cutting in the R&D process and acquisition system is essential to promoting projects that support the RMA.

Japan clearly faces a much different strategic situation than does the United States, and so the Japanese vision of the IT-RMA differs from the U.S. vision. The U.S. version is designed to conduct operations in foreign countries while Japan's SDF is primarily concerned with territorial defense. The next section discusses the history of the JDA's bid for the IT-RMA and Japan's visions for the future.

THE FUTURE DIRECTION OF THE IT-RMA IN JAPAN

A Brief History of Japan's Quest for the IT-RMA

Japan's steps toward the RMA have lagged far behind those of the United States. Although the *National Defense Program Outline in and after FY1996 (1996 NDPO)*[19] and the *Mid-Term Defense Program (FY1996-FY2000)*[20] refer to some RMA elements, those references were made without any intention of systematically pursuing the RMA. Since 1999, however, interest in the RMA within the SDF has greatly increased. In February 1999, the Office of Strategic Studies within the Defense Policy Division, Defense Policy Bureau, JDA, began a study of the RMA in the context of the SDF's transformation.

In July 2000, a directors general was assigned to make SDF policy for the information revolution. His office published *Information Technology Revolution of JDA/SDF* in August 2000. Three key words framed the policy for an information revolution within the SDF: "quality" (establishing an advanced information network); "utility" (improving the performance of C4ISR); and "security" (keeping the network secure). In September 2000, the Office of Strategic Studies issued *Info-RMA: Study on Info-RMA and the Future of the Self-Defense Forces*,[21] which enumerated seven principles that would serve as guidelines for Japan's IT-RMA defense plan: information, jointness, quickness, efficiency, flexibility, protection, and interoperability.

While all the countries that seek the IT-RMA may share all of these principles, some reflect Japan's unique defense policy. "Information" is a fundamental pillar of the IT-RMA, but its application to Japan's defense posture is different from those of other countries. Because Japan's defense policy is based on EDOP, the SDF's main operational area will be limited to those areas surrounding Japan. The SDF does not need to develop a global sensor and communication network, but can instead concentrate its efforts on an advanced information system to be used within Japanese territory.

"Efficiency" also reflects Japan's unique defense policy.[22] This "efficiency" principle includes efficient movement of units and efficient use of weapons. Under EDOP, the SDF will have to intercept invading forces on Japanese territory so the need to limit collateral damage is extremely high. If the SDF has a highly sophisticated sensor system and advanced PGMs, it can efficiently destroy enemy forces with minimum collateral damage against Japan's social infrastructure. The "efficiency" principle means something different for the United States, which conducts military operations almost exclusively in foreign countries.

Based on these seven principles, *Info-RMA: Study on Info-RMA and the Future of the Self-Defense Forces* summarizes the goal of a post-RMA SDF as follows:

> Sharing real-time information among each unit of the Ground, Maritime, and Air-Self Defense Forces based on redundant and invulnerable information networks comprised of various sensors; securing interoperability between SDF

and U.S. forces; and establishing a defense posture that could perform most efficiently with a minimum of reaction time, and could respond flexibly in accordance with rapidly changing situations.[23]

This report only lays out concepts and principles for Japan's IT-RMA. *Guidelines for Comprehensive Measures for Information Technology Revolution of the JDA/SDF* was issued in October 2000 and describes battlefield-related concepts as well as RMA requirements for administrative affairs, acquisition systems, human and technological infrastructure, and technological cooperation with other countries. It explains Japan's plan for building an advanced information network and melding the separate RMA visions of each service of the SDF: the Ground Self-Defense Force's G-Net, which links information from a unit in the field to the General Staff Office; the Maritime Self-Defense Force's *Maritime Operation Force System* and *Command and Control Terminal*, which enhance information sharing capabilities among vessels at sea; and the Air Self-Defense Force's upgraded BUDGE system, which is the command and control system for the air defense of Japan.

In the beginning of 2001, the JDA started a new defense program called the *Mid-Term Defense Program (FY2001–05)*, a "shopping list" for the next five-year period that emphasizes information technology: integrating each service's network to build an advanced network; creating an advanced command and control system to promote information sharing from frontline units to headquarters in Tokyo; and strengthening information security to acquire the capability for cyber protection.

Outline of Japan's RMA

The core of the ongoing U.S. RMA is information-sharing and an intensive use of PGMs to deal with regional conflicts. U.S. programs can be summarized by rapid deployment, joint digitization of the command and control system, standoff PGMs, and stealth technology. One can especially see the change in air warfare by analyzing the Gulf War and Kosovo air campaigns. U.S. aircraft broke Iraq's and Yugoslavia's integrated air defense systems completely. As the United States demonstrated, a post-RMA force can easily infiltrate the opponent's air-defense system and destroy important targets, even those located far from the frontlines. The U.S.-led RMA is highly offense-oriented, optimal for operations in regional conflicts.

Offensive operations are not part of the SDF's mission so the Japanese application of the information RMA must be for defensive purposes. The pillars of Japan's "defensive" RMA should be digitization of the ground force, a joint information network, and development of tactical-level PGMs. Theater-level PGMs, like the Tomahawk cruise missile, have too long a range to be used in defensive operations.[24] There is little need for stealth technology because the main mission of the ASDF is not to attack ground targets but to intercept invaders' airplanes. ASDF defense planners will continue to place emphasis on fighters' climbing power and maneuverability rather than on stealth technology. Moreover, a rapid deployment capability is not as important for the SDF, whose primary mission is territorial defense. Unless Japan changes its defense policy and actively starts to dispatch ground troops for combat, there is little need for a "global reach" capability. For the current GSDF, firepower and armor are still more important than strategic mobility. Although heavy weapons like the Crusader howitzer are not suitable for the United States, they may still be useful for Japan because its forces are not concerned with power projection.

The Japanese version of IT-RMA has two additional characteristics: interoperability with the United States and BMD. Many problems of interoperability between the U.S. and its NATO allies occurred during the 1999 Kosovo air war,[25] including incompatibility of secure information, precision engagement capabilities, and intelligence gathering capabilities. Will such problems appear in the Japan–U.S. alliance?

The context of interoperability problems in the Japan–U.S. alliance are different from those that occurred within NATO. The Japan–U.S. alliance does not engage in the type of multinational integrated operations that NATO conducts. In the NATO case, the NATO Supreme Allied Commander Europe (SACEUR) has authority to command and control all units in the area of responsibility. Units from different countries take part in the same mission and in the same area, so a high level of interoperability is absolutely necessary.

By contrast, according to the current structure for roles and missions sharing between the SDF and U.S. forces, the SDF would implement defensive operations and U.S. forces conduct offensive operations. The forces do not conduct the same mission in the same area under a unified command structure so the problem of interoperability on the same battlefield is not necessarily a serious problem for Japan–U.S. defense cooperation, unlike NATO. However, the 1998 Defense Guidelines states that Japan will provide logistical support to U.S forces in case of military contingencies in Japan and surrounding areas. Without a digitization of logistics forces that is compatible with the U.S. system, Japan cannot provide effective logistical support to the United States. Digitization of logistics must be given high priority in Japan's IT-RMA.

Another dimension of Japan's RMA will be defense against ballistic and cruise missiles. After the 1998 North Korean *Taepo Dong* ballistic missile launch, the JDA decided to begin cooperative BMD research with the United States. In addition to the ballistic missile threat from North Korea, proliferation of weapons of mass destruction and ballistic missiles poses serious threats for the international community.

It is worth exploring the implication of BMD for the IT-RMA. The essence of the RMA is information sharing through advanced technology and the systematic use of sophisticated high-technology weaponry. Gathering information from sensors such as satellites, sharing it among units through high-speed information networks, and advanced "hit-to-kill" interceptors are critical components of BMD. The essential elements of BMD are identical to those of the information-based RMA. Indeed, BMD is a part of the RMA. Considering the current strategic situation in Northeast Asia, BMD is more important than the digitization of ground units in Japan because some countries in the region have ballistic missiles while no country in the region has the capability for a successful amphibious operation against Japan.

BMD operations are conducted under a special operational environment: detecting the trajectory of a ballistic missile is much easier than finding an enemy ground unit in natural terrain such as a forest. The process of distributing information about incoming ballistic missiles to an interceptor is easier than choosing which unit should launch a weapon against an opponent on a complicated battlefield. BMD can be the first step toward achieving the IT-RMA. By promoting BMD, the SDF will be able to acquire a critical element of the RMA and once deployed, learn how to overcome obstacles to digitization of other kinds of operations.

The threat of cruise missiles will also be very real. If the information-based RMA proliferates, regional countries that are enjoying economic development will deploy land-attack cruise missiles. The SDF must acquire countermeasures. Currently, it is difficult to detect a cruise missile because it flies at a very low altitude. But once detected, it can be shot down with existing technology and equipment. The information capability to detect is critical for cruise missile defense, particularly advanced sensors that can discriminate cruise missiles from ground clutter and information networks for transmitting target information to interceptors. An advanced digitized information network is indispensable for a cruise missile defense system. Both ballistic and cruise missile defense systems will be important pillars of Japan's IT-RMA.

Japan's Vision for the IT-RMA

There are two different paths for Japan's IT-RMA. The first path, a *full-scale RMA*, would be characterized by the SDF's promotion of digitization of all units of the three services and operational innovation for extensive and effective use of tactical PGMs. This vision would be in line with U.S. defense transformation, although Japan would not seek stealthy or standoff air strike capability and strategic mobility.

The second alternative is a *partial RMA*. The SDF would build a joint information network and seek operational innovation for the effective use of PGMs. Digitization would be limited to one brigade or division. Digitization of ASDF and MSDF would be promoted, while most parts of GSDF would retain their pre-RMA posture.

These two options share common elements. The first is logistical interoperability with the United States to provide support to U.S. forces in case of regional contingencies that may threaten the security of Japan, even if the situation does not involve a direct invasion of Japan. Failure to keep up with the digitization of the American logistics system would erode the credibility of the alliance. *Creating a logistical support system that is interoperable with the United States* must be a top priority.

The second common element is missile defense, especially BMD, to cope with the regional ballistic missile threat. Regardless of Japan's path to the IT-RMA, a BMD system must have very high priority. Cruise missile defense may be needed in the future, but currently only a limited number of countries have land-attack cruise missiles.

A comparison of the full-scale RMA and the partial RMA reveals that a partial RMA is more suitable for Japan's current strategic requirements. Japan will be able to defend against pre-RMA invasion forces before they reach Japan's territory. It is unlikely that any country in the region will complete the IT-RMA in the next several decades, so there is no immediate strategic need to promote the full-scale option. Nor would Japan's current fiscal condition permit the huge increase in defense expenditures needed to implement this option.

The partial RMA option has two advantages. First, it is less costly. Second, the SDF can obtain the infrastructure for a full-scale RMA through pursuit of a partial-RMA. Once one brigade or division is digitized, the force can be expanded in the face of a changing strategic situation. The pre-RMA SDF can defend Japan against invading forces from surrounding countries because the SDF is superior to potential regional invaders. Given current circumstances, a partial-RMA is better than a full-scale RMA for retaining strategic and financial flexibility.

This analysis has assumed that Japan will keep its current self-restrained defense policy. If Japan changes its defense policy and actively dispatches troops overseas for international peacekeeping or regional conflicts, both the SDF's strategic outlook and the vision of Japan's IT-RMA will have to adapt.[26] Japan would then need strategic mobility, a global command and control capability, and digitized ground units that are completely interoperable with American forces.

CONCLUSION

The emerging RMA is one way to adapt to a highly fluid post–Cold War strategic environment. Under current strategic conditions, the United States is pursuing an "offensive" RMA, developing standoff and precision attack capabilities with highly advanced IT. Other countries, facing different strategic conditions, will choose alternate paths. Unless it is strategically relevant, even highly advanced weapon systems will not support national security.

A historical example is instructive here. Approximately six decades ago, Japan built the biggest battleships in history: *Yamato* and *Musashi*. They had 18-inch guns while other countries' battleships had 16-inch guns. Their armor was thicker than that of other countries' battleships. In the event of a surface battle between battleships, these two battleships could exert tremendously lethal force against their opponents. However, they made slight contributions to the war. For Japan, the core feature of the Pacific War was defense of sea lanes to Southeast Asia and protection of cargo ships from U.S. submarine attacks. The most relevant forces for Japan were not the most technologically advanced battleships, but ASW forces. If Japan had allocated more resources for ASW, its efforts to transport natural resources from Southeast Asia would have produced better results. Battleships were also not very effective during the Second World War because they were out of date after Japan's attacks on Pearl Harbor and on HMS *Prince of Wales* and *Repulse*. These revolutionary successes of Japanese naval aviation made their own battleships irrelevant to the new battlefield.

Two lessons follow from this history. First, every defense buildup must coincide with the strategic requirements of the country. Second, after an RMA, legacy forces face serious problems, even if those forces were highly capable on the pre-RMA battlefield.

To avoid the misfortune of *Yamato* and *Musashi*, Japanese defense planners must thoroughly analyze the current and future strategic contexts. They must build a vision, force structure and defense posture that are optimal solutions for Japan's strategic and political context. Copying the U.S. roadmap to the IT-RMA is not the answer. The SDF must keep pace with the ongoing transformation in warfare to avoid becoming a legacy force. Assuming that Japan's current defense policy will continue, a partial RMA based on interoperability with the United States in logistics and BMD is Japan's best solution.

NOTES

1. For Soviet views of the transformation of warfare, see: Mary C. FitzGerald, "The Soviet Image of Future War: The Impact of Desert Storm," in Willard C. Frank, Jr. and Philip S. Gillette, eds., *Soviet Military Doctrine from Lenin to Gorbachev 1915–1991* (Connecticut: Greenwood Press, 1992) pp. 363–386; Kimberly Marten Zisk, *Engaging the Enemy: Organization Theory and Soviet Military Innovation, 1945–1991* (New Jersey: Princeton University Press, 1993) pp. 120–177; Jacob W. Kipp, "The Labor of Sisyphus: Forecasting the Revolution in Military Affairs During Russia's Time of Troubles," in Thierry Gongora and Harald von Riekhoff, eds., *Toward a Revolution in Military Affairs?: Defense and Security at the Dawn of the Twenty-First Century* (Connecticut: Greenwood Press, 2000) pp. 87–104.
2. James R. Blaker, *Understanding the Revolution in Military Affairs: A Guide to America's 21st Century Defense*, Progressive Policy Institute Defense Working Paper, No. 3 (Washington, D.C.: Progressive Policy Institute, 1997) pp. 4–8; Theodor W. Galdi, "Revolution in Military Affairs?: Competing Concepts, Organizational Issues," *CRS Report for Congress 95–1170* (Washington, D.C.: Congressional Research Service, December 11, 1995) pp. 2–4.
3. See Joseph S. Nye, Jr. and William A. Owens, "America's Information Edge," *Foreign Affairs*, 75:2 (March/April 1996) pp. 20–35; Eliot A. Cohen, "A Revolution in Warfare," *Foreign Affairs*, 75:2 (March/April 1996) pp. 37–54; William A. Owens, *Lifting the Fog of War* (New York: Farrar, Straus and Giroux, 2000); Andrew F. Krepinevich, Jr., "The Military–Technical Revolution: A Preliminary Assessment" (Washington, D.C.: The Center for Strategic and Budgetary Assessments, 2002).
4. Clifford J. Rodgers analyzed the difference between RMA and Military Revolution from the perspective of military historian. See Clifford J. Rogers,

" 'Military Revolutions' and 'Revolutions in Military Affairs': A Historian's Perspective," in Thierry Gongora and Harald von Riekhoff, eds., *Toward Revolution in Military Affairs?: Defense and Security at the Dawn of the Twenty-First Century* (Westport: Greenwood Press, 2000) pp. 21–36.

5. See Clifford J. Rogers, ed., *The Military Revolution Debate: Readings on the Military Transformation of Early Modern Europe* (Boulder: Westview Press, 1995).

6. National Defense Panel, "Transforming Defense: National Security in the 21st Century" (December 1997) (www.dtic.mil/ndp).

7. William S. Cohen, *Report of the Secretary of Defense to the President and the Congress 1999* (Washington, D.C.: Government Printing Office, 1999) p. 121.

8. See Takahashi Sugio, "Joho-RMA and Kokubo Henkaku Kousou (Info-RMA and Defense Transformation in the U.S.)," in Kondo Shigekatsu and Umemoto Tetsuya, eds., *Bush Seikenn no Kokubou Seisaku* (The Defense Policy of the Bush Administration) (Tokyo: Japan Institute for International Affairs, 2002) pp. 135–162.

9. See Williamson Murray and Allan R. Millett, eds., *Military Innovation in the Interwar Period* (New York: Cambridge University Press, 1996).

10. See Geoffrey Till, "Adopting the Aircraft Carrier: The British, American and Japanese Case Studies," in Murray and Millett, eds., *Military Innovation in the Interwar Period*, pp. 191–226.

11. Colin S. Gray, "Nuclear Weapons and the Revolution in Military Affairs," in T. V. Paul, Richard J. Harknett, and James J. Wirtz, eds., *The Absolute Weapon Revisited* (Ann Arbor: The University of Michigan Press, 1998) p. 117.

12. Japanese defense policy is based on the article 9 of the Constitution. Article 9: (1) *Aspiring sincerely to an international peace based on justice and order, the Japanese people forever renounce war as a sovereign right of the nation and the threat or use of force as means of settling international disputes.* (2) *In order to accomplish the aim of the preceding paragraph, land, sea, and air forces, as well as other war potential, will never be maintained. The right of belligerency of the state will not be recognized.* This part of the Constitution does not forbid having armed forces for defense of the country. On the defense white paper of Japan explains the view of Japanese Government. See Defense of Japan (Tokyo: Urban Connections, 1999) p. 53.

13. The National Institute for Defense Studies, East Asian Strategic Review 2001(Tokyo: the National Institute for Defense Studies, 2001) pp. 308–311.

14. This "guideline" states the responsibility of SDF and USFJ as later. "The Self-Defense Forces will primarily conduct defensive operations in Japanese territory and its surrounding waters and airspace, while U.S. Forces support Self-Defense Forces' operations. U.S. Forces will also conduct operations to supplement the capabilities of the Self-Defense Forces." "Operations to supplement the capabilities" of SDF means to use strike capabilities. The full article, "The Guidelines for U.S.–Japan Defense Cooperation" is available at http://www.jda.go.jp/e/policy/f_work/sisin4_.htm.

15. Joint research and development with the United States is the exception, like FS-X project. Japanese defense industry is not permitted to participate in joint production, because the Diet has interpreted the production of parts for other countries' weapon systems as the "export of weapons."

16. "Defense News Top 100," *Defense News* (July 21, 2003) pp. 58–64.

17. On the "anti-military norm," see Peter J. Katzenstein, *Cultural Norms and National Security* (Ithaca, NY: Cornell University Press, 1996).
18. Ministry of Finance, "Heisei 15 Nendo Yosan Seihu An (Budget Request for FY2003)," (December 24, 2002) (http://www.mof.go.jp/jouhou/syukei/h15/h15top.htm).
19. In this document, some aspects of RMA are implied in the following excerpts:

> It is appropriate that Japan's defense capability be restructured, both in scale and functions, by streamlining, making it more efficient and compact, as well as enhancing necessary functions and making qualitative improvement to be able to effectively respond to variety of situations and simultaneously ensure the appropriate flexibility to smoothly deal with the development of the changing situations.
>
> Efforts will be made to enhance technical research and development that contributes to maintaining and improving the qualitative level of Japan's defense capability to keep up with technological advances.

20. In this document, some aspects of RMA are implied in the following excerpts:

> Concerning command, control and communication, to ensure command and control structures by which the central authority can send timely and appropriate directives from joint and integrated points of view, when the headquarters building of the Defense Agency is relocated, build up a New Central Command System (NCCS). In addition, continue to promote various measures such as establishment of the Integrated Defense Digital Network (IDDN), improvement of command and control capability, and utilization of communication via satellite.

21. Office of Strategic Studies, Defense Policy Division, Defense Policy Bureau, Japan Defense Agency, *Info-RMA: Study on Info-RMA and the Future of the Self-Defense Forces* (December 2000) (http://www.jda.go.jp/e/pab/rma/rma_e.pdf) (Japanese version was issued in September 2000).
22. Because of Japan's "anti-military norm," it is politically incorrect to use the word "high-lethality." So, "efficiency" was chosen.
23. Quoted from *Info-RMA*, p. 9.
24. But acquisition of Tomahawk itself is not banned by Constitution.
25. James P. Thomas, *The Military Challenges of Transatlantic Coalitions*, Adelphi Paper No. 333 (Oxford: Oxford University Press, 1998).
26. If the modernization of China's armed forces includes a rapid development of amphibious capabilities, and Sino-Japanese relations sour, the JDA may need to consider full-scale RMA.

5

LEARNING AND CATCHING UP: CHINA'S REVOLUTION IN MILITARY AFFAIRS INITIATIVE

You Ji

The Revolution in Military Affairs (RMA) has become the biggest military challenge for China in the twenty-first century. For Chinese political and military leaders, the RMA represents a new type of war of mass destruction, a new military theory, and a new historical phase of human development characterized by the endless innovation of high technology. The RMA is regarded as the major arena of great power competition that will decide China's position in the world over the next 20 years.[1] Jiang Zemin, China's outgoing commander-in-chief, defined the RMA as the product of a world economy moving from the industrial age into the information technology (IT) age. To advance with the times (*Yushi jiujin*), a famous political slogan of Jiang's, China has to revise its goal of national defense from seeking mechanization to hardware modernization and informatization *xinxihua*.[2] This constitutes the RMA with Chinese characteristics.

This chapter provides a Chinese perspective on the RMA. It explores how the People's Liberation Army (PLA) is responding to the global trend of IT-RMA-driven military transformation and analyzes the role of the RMA in the PLA's transformation. The chapter argues that if China can successfully translate RMA concepts into strategic guidelines, weapons programs, and force restructuring, given time the PLA will take on a new look and the entire Asia-Pacific region will feel the consequences.

EMBRACING THE REVOLUTION IN MILITARY AFFAIRS

The PLA has worked hard to understand the effects of the RMA on military establishments. In 1998, PLA National Defense University convened a major workshop on the RMA. Participants proposed a definition of the RMA as composed of revolutions in five areas: military thinking, military technology, military equipment, strategic theory, and force structure.[3] In the following years, especially since 2003, a comprehensive campaign to learn about the

RMA and deepen military reform has gained momentum. Nowhere outside the United States has the RMA been so seriously studied and translated into war preparation as it has been in China.

The Threat-driven Learning Campaign

An acute perception of external threat is the primary driver of PLA efforts to learn about the RMA. Under the pressure of Soviet invasion and the shadow of a potential Sino-U.S. conflict during the Cold War era, the PLA closely followed what was happening in these two militaries.[4] PLA researchers were aware of the concept of the RMA as early as the 1970s when they read articles about a revolution in military technological affairs in Soviet military journals.[5] PLA analysts have been sensitive to any sign of new developments by other major military powers. It has become a rule that as soon as something new is spotted, it should be thoroughly assessed and reported to the top command. The international discourse on the revolution in military technological affairs led to the PLA's systematic study in the 1980s of the linkage between technological advancements and military affairs. A consensus gradually emerged that a qualitative change in military science was in the making, a change brought about by the development of high-tech conventional weapons and information warfare (IW) systems.[6] In his keynote speech to the PLA's first all-Services conference on future war in 1986, General Zhang Zhen, then president of the PLA National Defense University and China's most influential military elder, warned that if the PLA did not take advantage of developmental trend in military science, it would be left behind.[7]

The relationship between a revolution in military science and its consequences for warfare rose to prominence in PLA thinking only in 1991 when Operation *Desert Storm* demonstrated how revolutionary technologies had transformed warfare.[8] The direct result was the PLA's adoption in 1993 of a new national defense strategy, "fighting limited wars under high-tech conditions," replacing Deng Xiaoping's strategy of "people's war under modern conditions."[9] Immediately afterwards, the Central Military Commission (CMC) launched a nationwide campaign to study the American RMA in order to learn how the U.S. military planned to fight at the turn of the century. Since the mid-1990s, the concept of the RMA has attracted enormous interest among the rank and file of the PLA.

The motivation for the PLA to learn from major military powers is threat-driven. China's assessment of the post–Cold War world order has been quite pessimistic.[10] Deng's assertion that major wars would not erupt has disappeared from PLA writings. Jiang has stated repeatedly that negative new trends in international politics have imposed great urgency on China's military and technological modernization efforts.[11] In November 1999, soon after the attack on the Chinese embassy in 1998, the Politburo made a decision to accelerate military preparation for war.[12] The American RMA has been eagerly studied in the PLA because the United States is perceived as the only country that the PLA may have to fight in the future. China's strategists have not failed to notice that the U.S. military has also seriously studied how to fight the PLA.[13] Post-Tiananmen sanctions, forced inspection of the *Yinghe* shipment, bombing of the Chinese embassy, the EP-3 incident and President Bush's pledge to defend Taiwan all have convinced the Chinese where the threat to their interests lies. The United States is seen to be using the RMA to consolidate its superiority over other world powers, especially over China.[14]

The PLA is also aware that China's neighbors are modernizing their militaries along RMA lines, and many are China's potential opponents. Japan and Australia may face requests from the United States to assist in a Sino-U.S. war over Taiwan. India's military modernization is increasingly guided by the RMA. Taiwan's military buildup is clearly RMA-driven. China's traditional defense culture stresses the importance of knowing the enemy. These regional developments accompanied by rising fears of superior U.S. power guide the search for a suitable model for PLA reform.

The RMA represents a grave challenge to the Chinese military because the PLA will have limited capabilities to deal with high-tech wars for the foreseeable future. Any breakthroughs in military technology will occur slowly and China may be left further behind. The PLA's eagerness to learn and absorb the RMA reflects China's realistic assessment of potential future conflicts[15] and China's time-honored mentality of seeing national security primarily in terms of credible military power. The RMA may be an opportunity for China to catch up with its military competitors.

Learning and Understanding the RMA

The PLA sees the RMA as a larger and deeper revolution in social and economic development that is reshaping the way people live and think. According to Professor Zhu Guangya, China's top defense scientist, the RMA is the product of socioeconomic and technological developments in the post–industrial age. The combination of advanced weapons systems, new military theoretical guidelines, and suitable force structure can generate a qualitative improvement in the employment of military power. However, without a full grasp of the deeper political and economic underpinnings of the RMA, it will be difficult to work out the timing and correct framework for combining these three crucial factors.[16] The great test for China is whether it can continue to reform its political, social, and economic systems in order to free its people to innovate. This will decide the country's future.[17]

The PLA has advanced a number of principles to guide its learning process to correctly understand what the RMA means for China's military modernization. The first is emancipation of minds, which means combating conservatism (fear of change), departmentalism (vested bureaucratic interests), and empiricism (worship of old experience).[18] China is fortunate to have been exposed to the RMA at a time when its leadership faces minimal ideological constraints. PLA researchers argue that new technology will not win wars without new combat theory. New theory will not be invented without fundamental changes in the mentality of PLA soldiers.[19] The RMA stimulates the PLA to shake off its historical burdens rooted in revolutionary ideology and old military strategies.

The second principle guiding the learning process is that of closely following the new theories and practices of the major military powers. This is regarded as crucial to implementing the instructions of Jiang to win the next high-tech war.[20] The PLA believes that recent limited high-tech wars provide good cases for China to study in order to understand the logic, operational features, and combat patterns of its potential adversaries, and to design countermeasures against sustained air attack or aircraft carriers. PLA researchers regard the demonstration effect of the American RMA in recent wars as one of three fundamental factors helping China to innovate its own combat theories and principles.[21]

The third principle guiding the learning process is that IW should be carefully studied to learn its merits and weaknesses. This is crucial for the PLA, which, for a long time to come, will have to rely on inferior weapons to fight more powerful enemies. PLA research institutions have studied the Kosovo war thoroughly: how NATO air attacks were hampered by bad weather and difficult terrain; why Yugoslavia's integrated air defense system could not shoot down sufficient numbers of enemy aircraft; why NATO forces failed to inflict heavy casualties on Yugoslavia's army; and the lessons the PLA should draw from this one-sided war which may mirror the situation the PLA will confront in the future.[22] The overall goal is to find ways of waging anti-RMA warfare.

THE RMA AND INTERNAL DEBATE

Not all military leaders embrace RMA ideas. Various schools of thoughts exist in the PLA, debating each other over the best strategies for China's war preparation.[23] Although the top political leadership has tried hard to forge a consensus within the PLA, there are those who continue to believe that China's economic and technological situation does not provide a firm enough foundation for transforming the PLA.

Emphasis on Homeland Defense

Deng's doctrine of "people's war under modern conditions" continues to influence the PLA although this school of thought is no longer dominated by Maoist and Dengist ideas of "people's war." Rather, the focus is on homeland defense.[24] After the terrorist attacks of September 11, 2001, the PLA closely scrutinized anti-terrorism wars. Land warfare was determined to be the key to any military operation designed to produce regime change, with air and naval power playing supporting roles, and guerrilla warfare inflicting heavy casualties on a casualty-sensitive invader.[25] This school of thought argues that the PLA should adopt a strategically defensive posture toward power projection beyond China's borders. This school is comprised mostly of army commanders in China's inland military regions and is ideologically supported by survivors of the Long March. They believe that the PLA is too technologically inferior to pursue the RMA. In a war with a superpower, China has no choice but to use its people's power and its current equipment to transform the war into one of attrition.[26]

Few generals really believe it is possible for any power to occupy China, yet the fear of regime change, either by peaceful evolution or through international military intervention (in Tibet and Xinjiang), resonates with a siege mentality and worst case scenarios. U.S. defeat in Vietnam and the Soviet disgrace in Afghanistan showed that people's war is not without its logic when defending a continental country. The utility of people's war against a threat of land invasion is still accepted, even in this high-tech era. An adversary would still have to think twice about the effectiveness of conventional land and guerrilla warfare in the IT age. The quick elimination of the Taliban and Saddam Hussein may have called into question the relevance of this school of thought but the partial revival of resistance in Afghanistan and intensified urban guerrilla warfare in Iraq lends validity to these views. Still, the influence of Long Marchers continues to dwindle as does the belief that land war on Chinese territory is likely.

Limited War Under High-Tech Conditions

The majority of PLA generals belong to the school of "limited high-tech warfare." They believe China's future war will be fought against lightening air and missile surgical strikes or sustained air, missile, and electronic bombardment, as occurred in Kosovo in 1999.[27] People's power can do little under these circumstances. This school is led by powerful military elders such as Admiral Liu Huaqing, and Generals Zhang Wannian and Chi Haotien, all former executive vice chairmen of the CMC (Liu retired in 1997 and Zhang and Chi in 2003). Liu, a powerful Long Marcher, strongly emphasized the application of high-tech weapon systems in modern warfare and placed key emphasis on equipping the PLA with as much high-tech hardware as possible. Nevertheless, his thinking was dominated by industrial-age ideas. He believed it was crucial for the PLA to complete the transition to mechanization before moving on to informatization[28] and advocated creation of heavily equipped group armies and aircraft carriers.

Generals Zhang and Chi were not opposed to using the RMA as a guide for PLA modernization yet as veterans of traditional wars, they are heavily influenced by the ideas of conventional ground warfare (two-dimensional war in the air and on land). They believe China must be realistic about its underdeveloped IT industries. Premature stress on developing systems of systems may slow the transition toward mechanization without making major inroads toward informatization. They prefer that China adopt individual aspects of informatization rather than try to fully integrate informatization and mechanization.[29]

Their view is similar to predictions made by U.S. analysts that China will be able to launch an RMA only after the second decade of the twenty-first century,[30] when it has developed sophisticated IT industries. Transformation in the PLA should occur through generational evolution, not a generational leap forward. During their command of the PLA, they promoted large numbers of their followers to key PLA positions across the four Services so this school represents the largest contingent in the PLA's officer corps.

The limited high-tech warfare school differs from the RMA school discussed in the following section only on the timing of the RMA and on how decisive informatization will be in future wars. According to this school, victory is secured on the ground, not through IW. They agree China needs to reduce its forces and restructure the PLA to develop the specialized Services like the air force and navy,[31] but there is no urgent need for a fundamental overhaul. Digital divisions and constructing a digitalized battlefield is not immediately relevant to China's situation. They believe it is too early to replace the current vertical C4I system that links the CMC, seven military regions, and a number of war zones with a new integrated five-dimensional C4I system (land, sea, air, space, and electronic space). Such a system would need to be horizontally distributed and would remove the existing functional divisions of command between geographical locations and the four Services. Maintaining the current force structure serves their interests, while any major overhaul would reduce their power, budget, and promotion opportunities.

The RMA as the Guiding Principle for PLA Modernization

To the RMA school, information power is as important as firepower.[32] Dominance of information will be a crucial factor in deciding the outcome of future war. In the PLA, the number of true believers in the RMA is small. Many simply mimic their U.S. colleagues, and lack a deep understanding of the RMA.[33] The enthusiastic supporters of the RMA are the war planners in PLA headquarters who are very familiar with U.S. armed forces and the academic staff in PLA educational institutions who train future PLA leaders. They have spearheaded the study of the RMA and worked hard to convince PLA senior officers that times have changed. They are young, well read, visionary, and anxious to create a new PLA that is more professional than revolutionary.[34] They support China's modernization but reject its wholesale Westernization. They entertain strong nationalist feelings but oppose xenophobia. They see communism as irrelevant to China's goal of self-strengthening but accept the Chinese Communist Party (CCP) as the vehicle for the realization of that national goal. They are painfully conscious of their country's present state of military backwardness. In the years to come they will wield increasingly more influence in the PLA due to their strategic positions in PLA headquarters and their personal networks forged with field commanders in PLA tertiary educational institutions.

Their views on the RMA are shared by the top civilian leadership. Jiang is a sincere RMA advocate. Although semi-retired, Jiang is still well regarded by the most senior members of the military who owe their promotions to Jiang's personal support. Jiang was successful in placing a number of generals who are committed to the RMA in the new CMC at the 16th Party National Congress in November 2002. Jiang's RMA ideas have been adopted by his successor Hu Jintao, the current party boss, state president, and deputy commander-in-chief. In a rare public speech on military affairs in May 2003, Hu Jintao called on the PLA to embrace the RMA and deepen PLA reforms.[35]

This school sees information as the heart of the RMA. Victory in a twenty-first-century war will no longer depend on the number of tanks and warships but rather on a country's computers, satellites, and software.[36] According to Major General Chen Youyuan, director of the Officers Training Bureau in the General Staff Department, the RMA is changing the basic structure of the military and the relations of the Services to each other; introducing new campaign formations and means of engagement; creating a much larger combat space embracing five dimensions (land, air, sea, space, and cyberspace); creating new modes of operations centered on IW; and creating new methods of combat: indirect, non-positional, and invisible.[37]

This school stresses the role played by systems of systems and how informatization can multiply the capability of conventional arms. Thinking in simplistic terms of adding numbers of hardware platforms to improve military capabilities is obsolete.[38] The PLA must accept several principles for its long-term modernization:

Strike From a Long Distance. New sophisticated terminal guidance systems and precision weapons have made possible attack beyond visual range. This will minimize human engagement and greatly reduce casualties. PLA researchers have studied U.S. concepts of combat "disengagement and indirect assault" and "concentrated firepower but dispersed manpower," and accept U.S. claims that in the future tank battles, aircraft "dog-fights," and exchange of fire by big naval guns will be history.[39]

Small-Sized Battle Formations Without Compromising Strength and Outcome. Small elite forces and simple-layer C4ISR systems are more suitable for IW, which is characterized more by combat between hardware/software than between men. With digitized and precision ammunition, a small high-tech force can overpower an army ten times larger. Digitization provides a high level of battle-field transparency to the side that possesses the means for multidimensional intelligence acquisition.[40]

Linkage Between Superiority in Information and Victory of an Operation. IT is not only an indispensable means for better command, control, and communication, but also constitutes an effective weapon to kill the enemy directly by attacking their ability to gather, process, and analyze information with computer virus attacks and software bombs. Superiority in information technology leads to superiority in combat operations.[41]

To its advocates, the RMA is a political requirement, military imperative, and matter of national survival.[42] Internal debate about the RMA exists, but there is a general consensus that the information age will profoundly affect war preparation, and that technological breakthroughs have visibly altered the conduct of war. All PLA personnel agree that China needs to catch up with these developments but useful elements from each school continue to guide strategic planning. The idea of people's war is valid in a conventional war against an invading enemy. The high-tech war school is valid for dealing with the main type of action China envisages in the near term: limited armed conflicts conducted with high-tech weapons. When China has made significant technological progress, the PLA will be able to implement RMA ideas, using advanced military satellites, miniaturized super computers, and IT networks to digitalize its armed forces. The RMA is more a theoretical blueprint than a concrete roadmap for now. Increasingly, the RMA school is unifying the thinking of the PLA high command.[43]

CONSTRUCTING A CHINESE MODEL OF THE RMA

The RMA is still in its formative years. Its initial phase is expected to extend to 2020,[44] giving the PLA time to design a "catch-up" strategy. The PLA believes that the application of RMA ideas and practices must suit China's strategically defensive posture, for example by developing tactics for asymmetric warfare. In the view of PLA researchers, the American RMA is designed to maximize offensive capabilities in a global setting. This does not fit with China's traditions, current practices, or future needs.[45] The key to sinifying the RMA is for the inferior military power to realize that it cannot mechanically copy the RMA force structure and combat patterns of the United States. The RMA is not a technological privilege that only a superpower can exploit.[46]

The Generational Leap Forward Strategy

The PLA is implementing its own version of the RMA,[47] built upon two simultaneous transformations: mechanization and informatization. For Western militaries, RMA transformation followed after a high level of mechanization had been achieved. China's RMA strategy is a "generational leap" strategy, moving toward informatization even when mechanization is far from complete. This "double construction" is the primary mission of PLA modernization, set in motion by Jiang at the end of 2002.

Jiang defined this strategic initiative as "using informatization to upgrade mechanization and using mechanization to accelerate informatization." Mechanization will not proceed along a traditional path of adding more conventional hardware such as tanks. As the two processes move forward, IT assets will be plugged into all weapon systems so that mechanization occurs more quickly and effectively.[48] Double construction is not a political slogan but represents a practical approach to military engineering. This strategy is an unconventional guide for PLA transformation, based on a realistic assessment of RMA trends in the world and the PLA's own particular situation.[49]

The leap forward strategy is risky. Three factors may hinder its realization. First, a weak national economy and IT sector, cannot sustain comprehensive military modernization in the near future. Second, China's IT industries are still in the early stages of development and key technologies must be imported. Third, serious problems exist in the PLA's equipment system and training system, and painful reforms will take time.[50] One crucial precondition for the strategy to succeed is correctly selecting priorities for military research and development (R&D). This is Jiang's principle of "do something and not others" (*Yousouwei yousuo buwei*), a rational choice given China's limited resources. The problem is setting priorities for investment. IT technologies, including R&D of IW systems, C4ISR facilities, early warning networks and electronic warfare architectures, are all top priorities for the PLA in its drive to achieve informatization. One key goal is to construct an integrated system of systems at the strategic and tactical levels. The development of space assets is regarded as the key to successfully leaping ahead a generation and has received a great deal of investment.[51]

Matching IT Force Multipliers with Low-Tech Equipment

The model of double construction provides concrete guidance for the PLA to modernize and transform. In the short run, the PLA plans to network existing hardware equipment with available information platforms that will act as force multipliers making older weapons systems far more effective.[52] No matter how limited China's satellite technologies may be, wise use of them can improve the PLA's C4ISR interconnectivity, which provides real-time information to field officers and increases battlefield transparency. The 38th Group Army is a successful example of transformation based on this model. As the first mechanized unit, it is highly mechanized but backward IT systems seriously undermine its full firepower potential and mobility. Data must still be sent to the anti-air missile brigade manually. Reconnaissance units can collect intelligence quite effectively but when they penetrate to the "enemy's rear," they cannot transfer information back to command headquarters because they lack advanced telecommunications equipment. In recent years, the Army has substantially improved its IT capabilities to build a force structure for the IT age.[53]

Another example shows how PLA commanders plan to apply RMA concepts to China's force modernization. The PLA is currently organizing its first experimental digital brigade. The Chinese have studied how U.S. and Australian special forces operated behind enemy lines with zero casualties during the Iraqi war in 2003. Each soldier of the allied forces was provided with real-time information of the enemies' location. This high-level, one-way battlefield transparency characterizes the digital division. For the PLA, it was clear from the beginning that the regional command would not need to

know the location of individual soldiers, only of companies. This requirement substantially reduced the pressure on both the sophistication and avail-
ability of equipment. The goal is to use IT force multipliers to greatly
enhance the combat effectiveness of low-tech forces. At this stage, the RMA
for China is not about acquiring long-range precision weapons and global
strike capabilities that are the core component of the U.S. RMA, but rather
to explore "Chinese" ways to protect China against remote-controlled
attack, and to wage counterattacks against the enemy's homeland through
means like cyber warfare.

The RMA as the Core of the National Defense Strategy

The PLA has embraced the RMA as a guideline to improve its national
defense strategy, and is attempting to sinify the RMA to meet China's unique
defense requirements. The evolution of China's national defense strategy
over the last 30 years has been filled with doctrinal flaws and maladjustments
to the world situation. The challenge is to formulate a sensible strategy that
will stand the test of time. By the beginning of the 1990s, PLA generals realized that Deng's "people's war under modern conditions," which had been
China's official national defense strategy since the beginning of the 1980s,
provided a poor guide for the PLA's development. It conflated two very
different doctrines: relying on population power (war of attrition in the
heartland) on the one hand, and firepower (professional soldiers stopping an
enemy advance on key fronts) on the other. Deng's strategy was designed to
withhold a Soviet land attack through positional warfare, but this was out of
step with the evolution of international affairs soon after it was approved.[54]
The PLA's practical strategy during the second half of the 1980s was aimed
at fighting a limited regional war to deal with China's border disputes, but
this was not a long-term strategy. China, as a major world power, could not
base its military modernization simply on considerations of potential
conflicts in the South China Sea or along its borders.[55]

For a time at the turn of the 1990s, China experienced a vacuum in its
national defense strategy. The Gulf War supplied the Chinese with a concrete
image of future wars the PLA was likely to face. The Chinese leadership
promptly responded to the new situation. During his inspection tour to the
PLA University of Science and Technology in 1991, Jiang summarized his
understanding of future wars: these would not be protracted ones on a world
scale (or wars of attrition) but intensive high-tech wars fought on multiple
dimensions, wars of electronics and missiles. In late 1992, the CMC
advanced a new strategy, *fighting a limited war under high-tech conditions*,
which laid the groundwork for force restructuring, weapons development,
general training, and the formulation of "war games" to counter China's
potential threats.[56]

The high-tech strategy emphasizes high-intensity joint operations based
on mobility, speed, and long range. Fought across all dimensions of the

battle space simultaneously, these wars are information intensive and critically dependent on C4ISR.[57] The upshot has been several changes in China's national defense strategy based on its learning about the RMA.

First, the RMA calls for linking active defense with forward defense, which may mean power projection beyond the country's land borders. This represents a radical departure from Deng's active defense, which was confined to territorial defense in the form of positional warfare around major cities. Forward defense is key to the new strategy. It recognizes that in a high-tech war, the enemy can strike from a great distance and that the dividing line between the front and rear has become academic. The PLA greatly enlarged its strategic depth, which, according to PLA war planners, should not be confined within China's borders.[58] Although this may not prevent the enemy's long-range attack as demonstrated in Kosovo, effectively engaging the enemy in the outer defense line of the country may pose a greater threat to the enemy, improve the PLA's early warning, and reduce China's personnel and material losses.[59]

Second, the RMA is designed to maximize offensive effects. The digitized battlefield, electronic soft kill, and pinpoint elimination of the enemy's key targets all indicate that the side that seizes the offensive first has the best chance of success. An offensive posture and preemptive strike capability are especially crucial for a weak military like the PLA at the beginning of a high-tech war.[60] This belief has been fully incorporated into the PLA's high-tech defense strategy, producing an offensive-oriented strategy that reverses the doctrinal weaknesses of Deng's strategy. PLA strategists argue that China's post–Cold War military guidance should be changed from Deng's *yifang weizhu fangfan jiehe*, or "defense as overall posture, offense as the supplement," to *linghuo fanying gong fang jiehe*, or "adroit response based on a combination of offensive and defensive campaigns." Evolving high-tech hardware decisively favors a fast offensive strike.

Third, PLA military planning must be able to adapt to and absorb new technological innovations. The PLA's response to China's fluid security environment is a forward leaning strategy, focusing mainly on defense against the major powers, yet flexible enough to cater to different combat scenarios, from all-out high-tech warfare to small-scale border disputes. China's defense strategy is forward-looking also in that it is geared to preparing for war several decades from now. R&D priorities do not aim to equip the PLA in the next few years but rather ensuring it will be at the frontiers of high-tech breakthroughs in the future.[61]

The significance of future-oriented military thinking lies in China's belief that it is more important to establish the right direction for the PLA's future than it is to acquire advanced hardware.[62] Without sound theoretical guidance, even if the PLA acquires sophisticated weapons, it will not be able to use them to their full potential. RMA advocates always draw upon the same example to illustrate this point. In the 1930s, France and Germany possessed similar numbers of tanks. But French tanks were scattered across the army, while German tanks were concentrated in elite army divisions, making them unstoppable in ground battles. Different deployment methods produced vastly different effects in war, giving birth to a new RMA.[63]

Building a Favorable Environment for RMA Transformation

A consensus at the apex of power has made the RMA the guiding principle for PLA transformation. This is significant in and of itself. This political consensus exerts tremendous pressure on professional soldiers to adapt and over time will likely bring about a visible change in military culture by setting new criteria for officer training, selection, and promotion. Jiang made it clear to the PLA that his remaining usefulness was to press ahead with military reforms along RMA lines.[64]

A Pro RMA Culture in the Making

The RMA is not yet a PLA-wide consensus. It is just one of several options. Jiang's push for the RMA is therefore timely and specific. He is constructing a new pro-RMA culture in the PLA by reforming the PLA's personnel structure and promoting to leading posts generals who understand and are committed to RMA reforms. In the recent leadership reshuffle, Generals Guo Boxiong and Cao Guangchun (vice chairs of the CMC) have been put in charge of RMA-driven transformation. General Ge Zhenfeng, the new executive deputy chief of staff, is typical of the type of officer being promoted. He is famous for his deep knowledge of Western military science and wrote about General Patton when he was still a junior officer. In the rank of file, Jiang has initiated reforms in military training and education guided by the RMA. He has ordered the CMC to send large numbers of senior officers to study abroad and to recruit more post-graduates from top civilian universities.

Political support from Party headquarters has been crucial for this change of PLA culture. RMA advocates have a better chance of promotion with Jiang's blessing.[65] When these officers move into senior command posts, the PLA will be guided more visibly by RMA ideas. A consensus is in the making.

The change in PLA culture is key to its transformation, even though the change is still limited. Before technocrats took over the leadership in the CCP and in the military, the generalists-dominated leadership belittled the role of technology. As Mao remarked, nuclear bombs were paper tigers. Now, technocrats are making military decisions based on new technological innovations, discarding a long PLA tradition that identified strength with numbers of foot soldiers. As one senior PLA theorist commented, the inevitable RMA trend is that a military has to evolve away from drawing its physical strength from numbers (*tineng*) to relying on advanced technological hardware (*jineng*) and then further to becoming a military of intelligence (*zhineng*).[66]

Setting RMA Transformation on the Foundation of Civilian Technology

The RMA has become a driving force behind the development of science and technology in China. The reverse is also true. China's leadership has concluded that without a sound technological foundation, there is no point of talking about the RMA. Yet China's high-tech base is thin. Only in limited areas, such as the space industry, is Chinese technology competitive at the world level. Market reforms are gradually strengthening this weak foundation. China's rapid economic growth makes research in high-tech weapons more affordable, although it is too early for China to seriously contemplate quickly narrowing the technological gap with the West. China's technocrats-turned leaders have also made it state policy that guns will not be privileged above butter.[67] They see that America's technological race with the Soviet Union helped the United States to achieve a superior economic position in the post–Cold War era. They understand that defense-related high-tech drives scientific and technological revolutions, that the application of military information technology can be wideranging and profitable.

Development of the new economy is now China's top national priority,[68] particularly developing the IT sector. China has devoted enormous resources to its IT industries, which have grown annually by over 30 percent in the last decade and are now, in terms of size, the second largest in the world. This rate of growth is likely to continue for another decade. Breakthroughs have been registered in areas such as super-computing. The latest invention, Shuguang-3000, is capable of 403.2 billion cycles per-second (GHz). In 2001, China developed the first proprietary CPU and is now capable of mass-producing high-digit megabit memory chips with 0.25 micron widths. Defense-related R&D relies on technological innovations from China's fast-growing non-state high-tech firms that enjoy close ties to the world market. The PLA has become increasingly dependent upon the civilian sector for weapons upgrading and military modernization.

PUTTING RMA IDEAS INTO PRACTICE

China is in the early stages of transformation, making small steps toward translating the RMA into a force structure. Some of the initiatives underway are discussed here.

Initiating Qualitative Force Restructuring

The PLA has accelerated efforts to build a qualitative military by initiating large-scale force reductions. The PLA was ordered to downsize by 500,000 in 1997. When this round of streamlining was over in 2002, the Army had been reduced by 18.6 percent, the Air Force by 12.6 percent, the Navy by 11.4 percent, and the Strategic Missile Force (SMF), by 3 percent.[69] Over 1,500 administrative agencies at the army level in the central and regional command were cut. Two-hundred and ninety industrial and commercial entities and management bureaucracies were transferred to civilian government.[70] The size of the Army is the smallest since the PRC's founding. In 2003, another round of force reduction began to demobilize 200,000 men by 2005. More non-combat personnel will be streamlined, military education and research institutions will be merged, and most importantly, a number of group armies will be abolished.

Not only has the size of the Army been scaled down, but its structure has significantly changed. A major reform in the 24 group armies of the PLA, underway since 1999, has removed over a dozen divisions from the army structure. A large number of divisions and regiments have been downgraded to brigades and battalions. The guiding principle is in line with RMA thinking: a smaller force and simpler command chain. Technological strength has been upgraded with increases in units of missiles, aviation, communications, and electronic warfare.[71] The first infantry missile brigade was established in the Nanjing Military Region in 2001. The unit is equipped with highly accurate M-9 and M-11 ballistic missiles and specializes in the use of tactical conventional missiles in joint operations, a task that differs from the SMF. In contrast to a downsized Army, the specialized Services continue to enjoy priority for modernization. Enormous efforts have been made to consolidate second strike nuclear capabilities, create offensive air power, and develop a blue-water navy.[72]

Restructuring has occurred in other areas although progress is slow, to the point of annoying party leader Jiang.[73] The most visible change in the top command structure has been the establishment of a General Armament Department (GAD) immediately under the CMC in 1998, with the same rank as the GSD. This department has taken over the functions of weapons R&D, testing, acquisition, allocation, and related matters formerly assigned to various top agencies in PLA headquarters. The creation of the GAD is significant. First, it highlights the PLA's effort to accelerate the pace of developing advanced weapons systems, especially in IW. The GAD has been empowered to concentrate as many resources in this area as possible. Second, the GAD is important in implementing the RMA. The GAD ultimately will become the department of systems, coordinating informatization and integrating C4ISR of all the Services so the CMC can command the PLA more effectively. The department is currently in charge of China's military space programs and R&D of crucial IW facilities. It is laying a firm foundation for China to develop its own C4ISR network. Third, the GAD is in charge of weapons R&D and acquisition to meet the requirements of joint warfare. Weapons development by the different Services is currently uncoordinated so that weapons cannot be universally used by all the armed forces, weakening the PLA's capacity to conduct joint campaigns. The GAD oversees weapons programs of all the Services to ensure they serve the needs of future warfare.

Priorities in Weapons Development

RMA transformation is built upon advanced weapons systems, requiring the PLA to reset its weapons R&D priorities toward military satellites, IT facilities, fixed-energy and laser equipment, and electronic weapons. The PLA realizes that developing the requisite technologies to produce an indigenous land-attack cruise missile may be more important than the missile itself. The PLA's missile technologies such as propulsion systems (small turbojet engines and ramjet/scramjets) are relatively mature. The bigger challenge lies in the area of software development. R&D for guidance technologies (GPS for in-flight navigation and terrain contour matching guidance, infrared imaging, or synthetic aperture radar for terminal homing) has been slow and delayed the missile from becoming operational. The RMA has corrected China's long-held misconception that numbers and ranges of missiles matter more than software.[74] In addition to the hard work on guidance systems, China is also developing low observability technological capabilities for reduced radar crosssections and heat signatures.[75]

China's R&D of laser technology has advanced to the weaponization stage. The concept of the military use of space is also enthusically embraced.[76] Between 2005 and 2010, China's space-based surveillance architecture could have at least four components: (1) synthetic aperture radar satellites for all weather, day/night monitoring of military activities; (2) electronic reconnaissance satellites to detect electronic emissions in the Western Pacific; (3) mid-high resolution electro-optical satellites for early warning, targeting, and mission planning; and (4) a new generation of high-resolution recoverable satellites for intelligence and analysis.[77] The PLA believes that ultimately it will have to develop anti-satellite capabilities in order to deal with enemy's star wars initiative.[78] This effort will require continuing and sizeable increases in the military budget.[79]

Implementing New Campaign Tactics

Closely linked to the PLA's adoption of an RMA-related national defense strategy are efforts to develop campaign tactics in accordance with RMA-type warfare. At a PLA conference on campaign theory in late 1996, General Hu Changfa, deputy president of the PLA National Defence University, said that studying new campaign tactics and new patterns of engagement in campaigns in order to respond to new forms of warfare and new opponents had become China's primary task in order to win the next war.[80] At the conference, the PLA identified the two most likely forms of engagement in a future campaign: mobile and joint operations.[81] In recent publications, PLA commanders no longer separate these two forms, but refer to mobile joint operations. For some time to come, the PLA will deal with these two forms separately in order to develop the necessary hardware and software for each.

Mobile operations dictate a major revision of the PLA's operational doctrine that was centered on positional warfare to support a strategically defensive posture. New PLA campaign theory promotes offensive campaigns in mobility to support its strategically defensive posture. Defense and strikes in depth requires long-range mobility, and stifling attacks at the enemy's rear with precision missile and electronic bombardment. Inevitably campaign operations will have to be supported by IW measures.[82]

The essence of mobile operations is offensive-oriented (*gongshi zuozhan*). As PLA strategists point out, attack will be the main form of PLA campaign engagement, obviously with the Taiwan war scenario in mind.[83] Mobile operations are seen as key components of any campaign in the IT age. High-tech limited wars lack fixed borders and operational formats will change rapidly. Only through mobile operations can the PLA seize the initiative. China's strategic landscape also requires mobile operations. In future campaigns, the PLA may need to operate in multiple strategic directions and over a vast combat space. It will have to move rapidly in order to establish temporary geographic superiority *vis-à-vis* more powerful opponents.[84]

The PLA's new campaign tactics also focus on joint operations, which are seen as characteristic of RMA-led warfare, a departure from the PLA's traditional emphasis on combined operations for combat on the ground. "Combined" refers to employment of different arms of the services (*junzhong*) within the Army: units of tank, artillery, engineering, telecommunications, anti-chemical warfare, and others, brought together to execute a ground campaign. The specialized Services—navy, air force, and missile force—played only a minor role. The level of China's overall military technology prevents the specialized Services from providing close-in support to army units. China's armed forces grew up around, and are dominated by, the Ground Force, but the PLA high command realizes that in future, there will be no chance of victory if army operations are not supported by the specialized Services, especially by the air force.

The specialized Services have been directed to develop the doctrine and upgrade the hardware for joint operations with the Army. Their status has risen *vis-à-vis* the ground force given their indispensable role in any future war. According to Lieutinent. General Hu, landing operations will be the PLA's form of engagement and these must be joint operations.[85] In a joint landing exercise at the divisional level in East China in October 2001, the commanding officer was a naval officer, purposefully appointed as a vice chief of staff of the Group Army that constituted the main force for the exercise. He commanded all participating units from the navy, air force, and the missile force.[86] Although not seen frequently, it represents a future trend.

The RMA as a Guide for the Design of Future Wars

The RMA has broadened the PLA's vision in designing its guidelines for future wars. High-tech not only generates strength, but also exposes weaknesses of the superpower. The 9/11 tragedy reflects the vulnerability of a mighty nation to new kinds of war. The PLA is contemplating various types of asymmetrical warfare for self-protection.

Asymmetric Warfare: The Soft Kill of the Weak Against the Strong

The PLA envisages that IT warfare using computer viruses and cyber attacks to disrupt the enemy's command and control systems will be an integral part of twenty-first-century warfare. This will be much cheaper than attacks on aircraft carrier battle groups.[87] For the first time, militarily weaker powers have found the means to deliver punches to the soft-underbelly of the more powerful enemy. Waves of attack on websites in China and the United States following the EP-3 incident in April 2001 represented an embryonic form of such warfare that will become more widespread and intensified if a major conflict occurs.

The PLA plans to counter RMA warfare with IT warfare.[88] PLA strategists have a realistic assessment of China's vulnerability in a major maritime campaign so it is unlikely that the CMC will commit its naval fleets to an all-out sea battle with the Seventh Fleet. Rather, the PLA may choose a number of more cost-effective options such as electronic warfare. PLA strategists realize that the more a military depends on IT, the more vulnerable it becomes to attacks on its information hubs. One PLA researcher pointed out that about 80 percent of U.S. military communications facilities rely on civilian networks, creating a window of opportunity for cyber attacks.[89] The concept of "counter-RMA" provides a useful guide for thinking about how a weak military deals with a more powerful one. China is developing electromagnetic pulse weapons, laser guns, electronic jamming equipment, and computer viruses, and training computer hackers to attack enemy networks.[90]

Asymmetric Warfare: The Missile Threat

The PLA sees missile attack as another useful weapon for waging asymmetric warfare with a more powerful adversary. Saturated conventional missile attack is one of the first choices in a Taiwan war scenario, and explains why the PLA has accelerated missile deployment in Fujian. Missile attack is also a way the PLA can deal with aircraft carrier groups. The PLA does not currently possess this capability, but are determined to acquire it and are seriously studying the tactics. This war plan probably evolved from an alleged computer simulation by the U.S. Department of Defense that claimed the PLA had won a sea battle against the United States using concentrated missile attack.

The goal of missile attack, to bring the war to the enemy's territory, also constitutes an effective means of deterrence. Pinpoint missile strikes against the opponent's military bases, warships, and supply lines is less threatening politically and strategically than direct engagements between personnel. It reduces human losses for the PLA, and has a greater psychological effect on the opponent's population. Missile launches are also more manageable. They can be terminated promptly to avoid escalation and direct confrontation with the superpower. Yet concentrated use of missiles can paralyze carefully selected military targets in Taiwan.[91] Missile warfare is seen as the extension of a process for facilitating political settlement.

In this decade and the next the PLA will make it a top priority to enhance its missile capabilities. New generations of long-range and accurate missiles will be developed to compensate for China's inferior offensive capabilities. China's navy and air force remain weak, so conventional missiles are one of China's few deterrents against a major power.[92] Countermeasures are also being developed against the Theater Missile Defense systems that threaten to neutralize PLA missiles.[93]

Missile warfare is considered to be an indispensable part of joint operations. The missile force has traditionally pursued its own war doctrines and training programs. Joint campaigns seldom included the missile force. Technological improvements of conventional missiles have made the SMF a useful tactical offensive force that can join with other Services in likely war scenarios. For example, missile attack against an enemy's C4ISR centers and airfields assists air force efforts to achieve air superiority. The officers from the SMF are now required to join the headquarters of joint campaigns in each war zone, a significant departure from past practice.

OBSTACLES TO AN RMA-READY PLA

Even with a proper understanding of the RMA, PLA leaders still confront a tremendous task in applying this knowledge to the policy-making process. Political consensus at the apex of power does not mean automatic removal of systemic constraints, bureaucratic barriers, vested interests, budgetary limitations, weapons R&D and equipment priorities, and specific war plans. Together these pose major obstacles to a successful transformation of the PLA.

Systemic constraints stem from China's social and political system. China has demonstrated that with a huge investment and transfer of technology, a nation can achieve industrialization relatively easily. Yet in the IT age, a society must acquire "knowledge" power before it can transform its military. A closed authoritarian system stifles imagination and innovation, and imposes many taboos on political and academic discourse. China is slowly opening up. Its state structure is less rigid but social constraints are entrenched and will persist for a long time.

Vested bureaucratic interests adversely affect the RMA transformation process. An RMA-type military requires a simple, effective command structure. The PLA needs to overhaul its cumbersome organizational make-up by removing the layer of command immediately under the CMC. This layer, the seven military regions, serves administrative and operational functions, being in charge of field armies, garrison troops, and to some extent, specialized Services. Yet its administrative functions often confuse operational functions, resulting in delays of operational orders from above and combat information from below. Major General Wang Baocun described this structure as a vertical "tree type" that has long been obsolete. He believed that a key task of military revolution in China was to transform the vertical structure into one of horizontally distributed networks (wangzhuang).[94] This change is politically important as well. A "tree type" structure contributes to military regionalism, and Chinese history shows this, coupled with political and economic regionalism, can trigger national disintegration.

Even before the RMA was discussed in the mid-1980s, the PLA high command debated whether to transform each military region into a strategic area direction force whose headquarters would assume operational command only. This direction force, for example, Northeast China Strategic Force, composed of field armies, would no longer have administrative functions, such as the command over garrison troops. The CMC's relations with the area direction forces, their subordinate group armies and the basic campaign units (divisions/brigades) would become more direct. The proposed reform encountered strong resistance from senior officers who had support from powerful central commanders who used to serve in these military regions and maintained close ties with regional commanders. Opponents argued that military regions as a level of administration performed important political functions by securing social stability. Military regions were also in operational charge of the units of other Services in the area. Without this structure, it would be difficult for the four Services to be placed under a unified regional command, because the command of area direction forces fell to the ground force. These arguments played a key role in halting reform in the late 1980s.[95]

China's weak technological foundation represents an insurmountable obstacle for the PLA to launch substantial RMA reforms. Without available advanced technologies in telecommunications and new materials, a high-tech military has no foundation. Without a fully developed space industry, real-time command and control, reconnaissance and surveillance cannot be achieved. Precision strike is beyond reach. Manufacturing industries are also important for manufacturing high-tech equipment. China's scientists can design first-rate laser guns, missile navigation devices, and nuclear submarines, but its industries cannot manufacture them. If a military does not have enough fourth generation aircraft, integrated air defense systems and sophisticated naval ships, it cannot be considered an RMA force. Even if a military possesses all these platforms, without sophisticated systems of systems to interconnect them, it cannot wage high-tech war. China's ability to develop electronic equipment and software is even weaker. China's weak military technology is the result of its backward civilian technology and basic science, which for a long time to come, will obstruct the PLA's modernization efforts.

Limited funding poses another problem. An RMA-type military is very expensive. Despite China's record increase in defense spending over the last 15 years, there remains a huge gap between what is needed and what is available. According to China's science minister, the total national R&D expenditure was approximately US$12 million, or 1.1 percent of GDP.[96] Slow progress in the PLA's mechanization is primarily due to lack of funds. Up to this point, the PLA has been able to equip only two mechanized group armies (the 38th and 39th), much slower progress than planned. Each year Chinese civilian and military scientists produce weapons designs, but there is no money for testing. Those that do enter laboratory trials may be dropped later because there is no funding for production.[97] Available resources decide the pace, direction, and consequences of the RMA effort. Rising levels of unemployment and projects to develop industrial infrastructure force the

Party to direct investment away from the military. The PLA's RMA dream will not be realized for a long time to come.

Another hindrance to China's desired goal to build an RMA military is the make-up of the officer corps. Ninety percent of PLA officers today have tertiary qualifications.[98] While this figure is impressive, over half of these officers acquired their qualifications in nonmilitary and nontechnological areas. Their degrees are in personnel management, legal studies, political education, and public administration through correspondence courses and self-learning processes. These courses are easy to pass but are irrelevant to the RMA. Only 20 percent of PLA officers graduated from tertiary institutions of science and technology. Many officers are computer illiterates and ignorant about information systems, a fact that has hampered the PLA's new high-tech centered training program.[99] The majority of PLA soldiers are also from the countryside where basic education is inadequate. So a high percentage of soldiers cannot handle high-tech equipment and solve practical technological problems. This situation will likely endure for a long time.

The PLA must also deal with historical legacies such as the old mentality that people's initiative is more important than the power of hardware and information. Most of the PLA's victories were won on the ground, giving rise to a tendency to exaggerate the role of tactics and individual bravery, and downplay the role of advanced technology. PLA commanders now realize the danger of such a tendency when dealing with a powerful enemy that has good tactics and good hardware. Yet this problem has not been adequately addressed. First, given that the PLA's inventory will remain obsolete for a long time, it harms soldier moral if technological power is emphasized too much. Second, to the PLA, tactics is more important for a weak military when technology is unavailable,[100] although some PLA analysts point out that the focus of tactics is shifting from numerical strength to information superiority.[101] Dependence on tactics is related to another potentially harmful PLA legacy: inertia in its training programs. PLA basic units still place emphasis on night combat, based on the assumption that darkness provides good cover for a weak military. Yet darkness is actually an advantage for the military that possesses powerful night vision equipment.

None of these obstacles is insurmountable. Mental attitudes can change and historical legacies can wane with changes in the dominant culture of the military. Given time, a weak technological foundation can improve. China's high-tech civilian industries have grown at an unprecedented pace and will gradually deliver much needed technologies to the military. Shortfalls in the military budget can be addressed as the economy continues to grow. The biggest barrier is China's rigid political system with its vested interests. Whether China and the PLA can change in 20 years and accomplish their mission is anyone's guess.

CONCLUSION

The RMA has inspired the PLA to formulate its long-term modernization guidelines according to new rules of the game. To the PLA the RMA is the world standard, a developmental trend for a powerful military that it can ill afford to ignore. The very fact that China has so little capability to cope with the RMA type of war stimulates the PLA to study the RMA and to apply, where it can, its principles in practice. To China's leadership the danger of an RMA type of war against China lies less in its military consequences than in the political consequences that may follow: instability in China's social system and government. In the next few decades PLA watchers will see continuing reforms guided by the RMA. The PLA's C4I systems will gradually be streamlined and digitized. Its force size will be significantly trimmed and force components restructured to allow new specialized arms of the Services to emerge. Military R&D programs will emphasize the development of new weapons. National defense strategy, campaign tactics, and combat principles will be constantly reviewed based on the latest innovations of the major military powers. The PLA will gradually become more open, flexible, and forward leaning, more professional and high-tech oriented.

To the PLA rectifying doctrinal defects is more important in its long-term modernization than the immediate acquisition of modern combat hardware. Embracing the RMA and trying to put it into practice may be the right path for the PLA. Its official transitional model of double construction—mechanization and informatization—appears to be logical and rational. Yet it may underestimate the difficulties of both. In the end, it is very likely that mechanization will slow down while informatization does not deliver the goods for transformation. Adopting the correct strategic guideline does not guarantee success in transforming the PLA into a world class fighting force. The Chinese may be able to copy American thoughts but whether they can materialize the RMA is not clear.[102] If the Soviet failure in its technological race with the United States tells us anything, it is that a closed socio-political system may stifle the imagination of scientists and doom the long-term potential of the nation. The biggest challenge to China's search for major power status may not be its current technological backwardness but its rigid political system.

NOTES

1. Shengyue, "Yinjie xinjunshi gemin de sixiang tiaozhan" (Meeting the challenge of the RMA), *Huayue military forum*, August 13, 2003.
2. Interview with Major General Wang Baotong, *Program of National Defence*, China Central Radio, August 1, 2003.
3. Zhang Hui, "Xinjunshi gemin enti yanjiu xinlun" (The new ideas in the study of RMA), *The Journal of the PLA National Defence University*, nos. 2–3 (1998) p. 67.
4. The PLA has maintained large numbers of research institutes and think tanks earmarked for the study of major armed forces in the world since the early 1950s. Among the top agencies are the PLA Academy of Military Science, which has a large division on foreign militaries, and the PLA National Defence University, whose Institute of Strategic Studies has as its specific purpose tracking new defense strategies and theories in the United States and Russia. In each Service and military region, research institutes keep close watch on military innovations related to their modernization concerns. A famous one is the Navy's Centre for Equipment

Evaluation, established on the model of the Naval Analysis Corporation in the United States.

5. Cheng Bojiang, William Perry and others, "Junshi gemin yu meiguo kua shiji de guofang fazhan" (RMA and US defense development in the century beyond), *The Journal of PLA National Defence University* (December 1998) p. 50.

6. A large number of books and articles were published on the subject. An important and internally distributed book appeared in 1987 that recorded a top-level PLA conference on the revolution in military technological affairs in 1986. The contributors to the book were all senior PLA generals. Editors' group, ed., *Tongxiang shengli de tansou* (Exploring the ways toward victory), edited by a special research team, Beijing: the PLA Publishing House, 1987.

6. Liu Jishan and Qian Zunde, *Dangdai waiguo junshi sixiang* (Contemporary military ideas in foreign countries) (Beijing: the PLA Academy of Military Science Press, 1987) p. 32.

7. Zhang Zhen, "Guanyu wojun zhanyi lilun de jige wenti" (Some questions concerning PLA campaign theory), Editor Group, eds., *Tongxiang shengli de tansou* (Exploring the ways toward victory) (Beijing: the PLA Publishing House, 1987) p. 16. At the time he was president of the National Defense University and later the second in ranking in the PLA in the 1990s.

8. Su Yanrong, ed., *Xinshiqi guofang he jundui jianshe yanjiu* (Study of defense modernization and PLA building in the new era) (Beijing: The PLA Academy of Military Science Press, 1994).

9. You Ji's paper *China's Military Modernisation in the 1990s* to the workshop *China and the Asia-Pacific: Myths, Realities and Challenges,* Australian National University, December 2–3, 1993. It was later published in Stuart Harris and Gary Klintworth, eds., *China as a Great Power: Myths, Realities and Challenges in the Asia-Pacific Region* (New York and Melbourne: St Martin's Press & Longman, 1995) pp. 231–257.

10. Yan Xuetong from Qinhua University, for example, pointed out that the next 15 years would be most dangerous for China. Due to the Taiwan problem, a war may become inevitable. Yan Xuetong, "21 Shiji zhongguo jieqide guoji anquan huangjing" (The security environment for China's rise in the 21st century), *The Journal of the PLA National Defence University*, no. 1 (1998) p. 15.

11. Peng Ruiuan and others, p. 10.

12. General Qian Guoliang (commander of the Shenyang Area of Military Command) (2000), "Quanmian luoshi 'silinbu jianshe gangyao,' gaobiaozhun zhuahao silingbu jiguan jianshe" (Comprehensively implement the guideline of headquarters construction, and do a good job in headquarters construction), *Journal of the PLA National Defence University*, no. 6, p. 4.

13. See each of the annual reports on the PLA by the U.S. Department of Defense in recent years.

14. Li Jijun, "Xinjunshi gemin yu zhanlie siwei de biange" (RMA and changing strategic mentality), *The Journal of PLA National Defence University* (January 1999) p. 20.

15. As pointed out by Michael Pillsbury, the PLA's elaboration of the RMA has shaken up western notions about the backwardness of the PLA's strategic planning. Indeed, the very fact that the RMA is seriously studied in China is an indication of the PLA's development. Aside from the Chinese, only the Americans and Russians wrote on the subject in the early 1990s. Pillsbury, cited in *Far East Economic Review*, July 24, 1997.

16. Wang Gezhen and Li Mindtang, "Junshi geming souyi" (Some discussion on the revolution in military affairs), *The Journal of the PLA National Defence University*, no. 11 (1997) p. 22.
17. Gao Chunxiang and others, *Xinjunshi geminlun* (On the RMA) (Beijing: The PLA Academy of Military Science, 1996) chapter 2.
18. Cang Yetin, "Miaozhun junshi biange qianyan zhuangbian sixiang guannian" (Targeting the frontier of the RMA and changing our mentality accordingly), *The PLA Daily* (Internet edition), July 2, 2003.
19. Chen Youyuan, p. 38.
20. General Qian Guoliang, commander of Jinan Military Region, "Shiying tizhi bianzhi gaige xintedian yichuangxin jingshen tuijin budui zhiliang jianshe" (Conform to the new changes brought about by the structural adjustment and quicken the quality improvement with a creative spirit), *Journal of the PLA National Defence University*, no. 8 (1999) p. 33.
21. The other two are a protracted period of peace in the world and sustained domestic development, especially that of IT industries. "National Defence Forum," *China Central Radio*, August 10, 2003.
22. Huang Bi, "Kesuowo zhanzheng liang zhounian heigu" (Review of the Kosowo War after two years), *Journal of the PLA National Defence University*, no. 9 (2001) pp. 20–22. Chen Xianhua, deputy commander of Chengdu Military Region, "Chuangxin he fazhan gaojishu tiaojian xia renmin zhanzheng zhanlie zhanshu" (Innovate and develop people's war strategy and tactics under the hi-tech war conditions), *Journal of the PLA National Defence University*, no. 12 (2000) p. 21.
23. Gao Chunxiang, *Xinjunshi gemenlun* (On RMA) (Beijing: The PLA Academy of Military Science Press, 1997) p. 15.
24. Deng's doctrine of "people's war under modern conditions" prescribed that heavy infantry troops should be positioned along China's major cities. This was against Mao's strategy of luring enemies into the country's heart land.
25. Now this army-first tradition is regarded as a key obstacle for joint operations but the old sense still prevails. Yang Liuqing, "Shishi lianghe zuozhan ying jiejue de jige guanjian wenti" (Several key problems in waging joint operations that must be resolved), *The Journal of PLA National Defence University*, no. 8 (2002) p. 45.
26. Sun Jizhang, "Tantan zhanfa de texin" (On the pattern of war), *The Journal of PLA National Defence University*, no. 1 (1999) p. 47.
27. Hu Fan and others, *Tuyin zhizhan* (the fight between rabbit and eagle: comments on the Kosovo War by professor in the National Defence University) (Beijing: Zuanli wenxian chubanshe, 1999) chapter 8.
28. His strong push to build aircraft carrier is a typical example. See Ian Story and You Ji, "China's Aircraft Carrier Ambitions: Seeking Truth from Rumours," *The Naval War College Review*, December 2003.
29. Information gathered in China through interviews with PLA officers in 1999.
30. Liu Honji, "Lun youzhongguo tese de xinjunshi gemin" (On RMA with Chinese characteristics), *The Journal of the PLA National Defence University*, nos. 2–3 (1998) p. 62.
31. See for instance, Liu Yicang, *Gaojishu zhanzheng lun* (On high-tech war) (Beijing: Junshi kexue chubanshe, 1993) chapter 4.
32. Hao Chunxiang, *Xinjun geminlu* (On RMA) (Beijing: The PLA Academy of Military Science Press, 1997) p. 16.
33. *The PLA Daily*, December 17, 2002.

34. On this point, see You Ji, "China: From A Tool of Revolution to A Professional Military," in Muthiah Apalagapa, ed., *The Professionalism of Asian Armed Forces* (Hawaii, East-West Centre Press, 2001) pp. 93–110.

35. See Hu Jintao's speech on RMA to the enlarged Politburo meeting on The meeting was attended by all key PLA leaders. *The PLA Daily*, May 25, 2003.

36. Hu Yongfeng and others, *Shuzhihua budui yu zhanchang* (Digitalized troops and battlefield) (Beijing: Junyiwen chubanshe, 1998) p. 6.

37. Chen Youyuan, "Junshi jishu gemin yu zhanyi lilun de fazhan" (RMA and the development of campaign theory), *The Journal of PLA National Defence University* (January 1999) pp. 37–38.

38. *The PLA Daily*, August 9, 2003.

39. Chen Youyuan, p. 38.

40. Hu Yongfeng and others, *Shuzhihua budui yu zhanchang* (Digitalized troops and battlespace) (Beijing: Junyi chubanshe, 1998).

41. Li Qingshan, Xinjunshi gemin yu gaojishu zhanzhen (New revolution in military affairs and hi-tech warfare) (Beijing: The PLA Academy of Military Science Press, 1995) chapters 5 and 6.

42. Niu Qiwei and Xu Hong, "Jiaqiang guofan jianshe yao zhuyi senge shiying" (The three matching points in our effort to strengthen national defence), *The Journal of PLA National Defence University* (January 2001) p. 25.

43. You Ji, "The Supreme Leader and the Military," in Jonathan Unger, ed., *The Nature of Chinese Politics: from Mao to Jiang* (Armonk: ME Sharp, 2002) pp. 279–296.

44. Zhang Hui, p. 67.

45. As Emily Goldman and Tom Mahnken rightly pointed out, the idea of sinifying the RMA is a very intriguing one. They asked "when we think about how the RMA is being adapted abroad, should we think about 'Japanising' the RMA and 'Singaporising' it?" This is a good question. In my opinion, when the RMA is being adapted outside the United States, it will inevitably create new variations to suit local conditions. However, whether the new variations are significant enough to warrant such a definition as the indigenized RMA is very much open for debate. In most cases, it does not. There are American scholars who see the difference between "a general RMA" and "an American RMA." See for instance, John Battilega's paper to the conference on Conventional Arms Rivalry in the Asia-Pacific, October 24, 2001, Hawaii, organized by the Asia-Pacific Center for Security Studies.

46. Li Jijun, p. 22.

47. Liu Honji, p. 62.

48. Lu Xinagmin, "Dui xinxihua zhuangbei shixian kuayueshi fazhan de sikao" (Some thought on realizing generational leap in our information equipment), *The Journal of PLA National Defence University*, no. 6 (2002) pp. 84–85.

49. "Military Forum," *The PLA Daily on line*, July 12, 2003.

50. Lu Xinagmin, p. 84.

51. Wang Xingwang, "Fazhan taikong wuqi shi wojun zhuangbei jianshe kuayueshi fazhan de guanjian" (Developing space weapons is the key to the PLA's leap strategy of weapons equipoment), *The Journal of PLA National Defence University*, no. 10 (2002) p. 83.

52. "Jijiehua xinxihua tongbu fazhan wojun xiandaihua jianshe chengxiao xianzhu" (Simultaneous development of mechanization and informatization brings about a new look of PLA modernization), *Banyuetan* (Bi-monthly magazine), August 8, 2003, pp. 2–5.

53. *The PLA Daily*, August 11, 2003.

54. "Active defence strategy" was first raised by four star general Su Yu on January 11, 1979 to a group of senior PLA officer at the PLA Academy of Military Science. See, Su Yu, "Dui weilai fanqinlie zhanzheng chuqi zouzhan fangfa jige wenti de tantao", *Junshixueshu*, no. 1 (1979) pp. 1–12.

55. For more detailed analysis, see You Ji, *The Armed Forces of China* (Sydney: Allen & Unwin, 1999) chapters 2 and 7.

56. You Ji, *The Armed Forces of China*, chapter 2.

57. Council on Foreign Relations Report, *Chinese Military Power* (2003) p. 39.

58. Guo Yongjun, "Fangkong zuozhan ying shuli quanquyu zhengti fangkong de sixian" (Air defence should be guided by the theory of area and integrated defense), *Junshi xueshu*, no. 11 (1995) pp. 47–49.

59. Since the war against Vietnam in 1979, human casualties have become a sensitive political issue in China. Many people questioned whether it was worth the effort to attack Vietnam given so much loss of life. Although the traditional thinking that victory should be the ultimate consideration and in order to win, human sacrifice is necessary still holds in the military, the PLA is under increased pressure to minimize human losses.

60. Shi Zhigang, "Jiji fangyu zhanlie sixiang zhai xinshiqi junshi douzheng de tixian" (The application of active defense strategy in the military preparation in the new era), *The Journal of PLA National Defence University* (August–September 1998) p. 100.

61. Tao Bojun, "Dangde sandai lingdao jiti yu keji qianjun" (The Party's three generation leadership and strengthening the armed forces through technological breakthroughs), *China Military Science*, no. 3 (1997) pp. 65–73.

62. See articles in Pillsbury, Michael, ed., *Chinese Views of Future Warfare* (Washington, D.C.: U.S. National Defence University, 1997).

63. Xie Dajun, "Qiantan zhishi jingji jiqi dui junshi gemin de yingxiang yu tiaozhan" (The influence and challenge of knowledge economy to the RMA), *The Journal of PLA National Defence University* (January 1999) p. 27.

64. Interview with a senior officer from the PLA National Defense University in Sydney in May 2003.

65. They have written a large number of books and articles concerning the RMA, some of which are cited in this chapter. These articles are sometimes commissioned by senior officers for policy debate. Now all top generals have their "personal think tanks" whose primary task is to contribute new ideas. The RMA discussion is a major source of these ideas. Many of them have been incorporated into policy guidelines after a period of debate. Others are the reflections of the authors' reading of Western literature on the topic.

66. *The PLA Daily*, March 8, 1991.

67. Hu Jintao's speech to Politburo's study session of the RMA. *The PLA Daily*, May 25, 2003.

68. The Decision on Advancing Technological and Scientific Research by the Central Committee of the CCP in May 1995.

69. *Chinese Military Power*, p. 40.

70. Website of the New China News Agency, October 17, 2002.

71. Major General Liu Jiukuei (political commissar of a group army), "Nuli tansou xintizhixia jiceng jianshe de tedian he geliu" (Make a good effort to study the new features and patterns of the new army structure), *Journal of the PLA National Defence University*, no. 8 (2000) pp. 50–52.

72. Liu Zuoxin, "Kongzhong jingong zhanyi liliang goucheng he zhanyi zhihui chutan" (Initial research on the force structure and command for offensive air campaigns), *The Journal of the PLA National Defence University*, no. 10 (1995) p. 40.

73. Jiang has criticized the PLA many times for its slow pace of reform along the lines of the RMA since 1996 after he consolidated control over the military. Information from a senior PLA officer in May in Canberra 2003.

74. It is interesting to note that the PLA has spent more of its budget on satellite-based IT network than on building more missiles. Talk with researchers from the General Armament Department of the PLA in Beijing in December 2002.

75. Richard Bitzinger, "Going Places or Running in Place? China's Efforts to Leverage Advanced Technologies for Military Use," in Colonel Susan Puska, ed., *People's Liberation Army after Next* (Strategic Studies Institute, U.S. Army War College, 2000) p. 11.

76. China will launch over 30 satellites and a number of manned spacecraft in the next five years, according to Hu Hongfu, deputy general manager of the China Aerospace Industrial Group. *The People's Daily*, February 2001.

77. Mark Stokes, "China's Military Space and Conventional Theatre Missile Development: Implications for Security in the Taiwan Strait," in Colonel Susan Puska, ed., *People's Liberation Army after Next*, p. 113.

78. Cui Longzhu, "Qiantan fangkong zuozhan" (On anti-air operations), *Journal of PLA National Defence University*, no. 11 (1999) p. 53.

79. The increase in the military budget for 2001 was a record 17% over that for 2000. The trend will continue.

80. Hu Changfa, "Guanyu gaojishu tiaojian xia jubu zhanzhen zhanyi lilun yanjiu de jige wenti" (Several key questions of the theory of hi-tech campaigns), *Journal of the PLA National Defence University*, no. 1 (1997) p. 33.

81. Ibid., p. 34.

82. Wang Qunbo, "Jidong zhanzhong de kungjun zuozhan zhidao" (The guideline for air force mobile campaigns), *Junshixueshu*, no. 12 (1994) p. 28.

83. Hu Changfa, p. 34.

84. The remarks of General Liu Jingsong, president of the PLA Academy of Military Science, in his interview with Li Bin. Li Bin, "Dangdai shijie junshi fazhan qiushi jizhanlie qishi" (The developmental trend of the world military today and lessons learned from the studying the trend), *Journal of the PLA National Defence University*, no. 8 (1999) p. 27.

85. Hu Changfa, p. 34.

86. *The PLA Daily*, November 20, 2001.

87. *The PLA Daily*, April 20, 2003.

88. Thomas Christensnsen, "Posing Problems without Catching Up: China's Rise and Challenge for US Security Policy," *International Security*, 25 : 4 (2001) p. 25.

89. Lu Huilin, "Zhixinxiqun" (Counter information dominance), *Journal of the PLA National University*, no. 7 (1999) p. 35.

90. Christensnsen, p. 2.

91. Ibid.

92. You Ji, "Nuclear Power in the Post–Cold War Era: the Development of China's Nuclear Strategy," *Comparative Strategy*, 18:3 (August 1999) pp. 245–259.

93. The PLA will quietly increase the number of long-range delivery systems and warheads. It will also step up R&D efforts for multiple warheads reentry technology and deployment. This will be an area where an action-reaction arms race will intensify.

94. *People's Daily*, August 9, 2003.
95. Ian Wilson and You Ji, "Leadership by 'Lines': China's Unresolved Succession," *Problems of Communism*, 39 (January 1990) pp. 28–44.
96. Xu Guanhua, "Weilai zhongguo keji fazhan de zhanlie xuanze" (The options of China's strategies in developing science and technology), *Xinhua wenzhai*, no. 5 (2003) p. 145.
97. According to a senior military scientist, the ratio of the three categories (design, lab test, and production) is one-third each. He also informed me that China was able to launch manned-satellites a long time ago, say in the 1980s. Lack of funding delayed the project by almost two decades.
98. Among the PLA's 600,000 officers, 90% have higher education qualifications, 20,000 officers have Masters degrees and over 4,000 officers have Doctoral degrees. Wu Jianhua, "Wojun zhonggaoji nianqin zhihui ganbu baiyang de kaocha yu jishi" (A Review of the Promotion of Young Senior Officers in our Army and its Lessons), *Journal of the PLA National Defence University*, no. 1 (2000) p. 51.
99. Bao Chunliang, "Jiefangjun renyuan suzhi de bianhua," *Dongya jikan* (East Asia Quarterly) 31:2 (2000) p. 51.
100. Many articles on the so-called oriental tacticism have appeared in the Chinese military media. For instance, see *National Defence Daily*, August 5, 2003.
101. *Zhongguo guofangbao* (The newspaper of the national defence), December 3, 2001.
102. Ahrat, Ehsan, "Chinese Prove to Be Attentive Students of Information Warfare," *Jane's Intelligence Review* (October 1997) p. 17.

6

CHINA'S REVOLUTION IN MILITARY
AFFAIRS: RATTLING MAO'S ARMY

Andrew Nien-Dzu Yang

CHINA'S PERCEPTION OF THE REVOLUTION
IN MILITARY AFFAIRS

The term "revolution in military affairs" (RMA), literally translates into Chinese as "military revolution" (*jun shi ge ming*). The Chinese interpret this term differently than they do "revolution in military affairs." The Chinese understanding of the RMA includes at least three connotations. First, RMA implies a new "technological revolution" that has had a significant impact on the arms industry. The development, adoption, and integration of microelectronic technologies into weapons and platforms has created a revolution in the conduct of war characterized by the use of precision-guided weapons, information technology (IT)-based Command, Control, Communications, Computers, Intelligence, and Reconnaissance Systems (C4ISR), and revolutionary changes in operational concepts and tactics.[1]

Second, RMA is a "collection of systems" that uses IT to process information in order to yield real-time operational C4ISR, along with advances in doctrine and tactics to use these new capabilities in war.[2] Third, RMA refers to a new "type of military revolution" (*xin jun shi ge ming*), one that is entirely different from previous conventional technology-based military revolutions. It implies a fundamental change in weapon systems and equipment design, and is characterized by information-based precision-guided weapons, informatized operational platforms, digitized armed forces, and C4ISR. RMA also implies revolutionary approaches in military thinking and theories, as well as the structure of military organizations.[3]

China's armed forces, the People's Liberation Army (PLA), have been attracted to the RMA, but they have not entirely embraced, let alone undergone, a military revolution. Three obstacles prevent the PLA from pursuing the RMA. First, homeland defense still dictates China's overall military strategy, with sovereignty and territorial integrity posing primary national defense concerns. Ground force commanders dominate decision-making, leaving the air force and the navy as adjunct services. This may be a traditional feature of the PLA, but it has hindered the development of joint operations. Second, China's technological and industrial backwardness has prevented the PLA from upgrading its technology and producing state-of-the-art weapon systems and platforms, which is a "must" in pursuing the RMA. Third, this backwardness, combined with insufficient education and training, has slowed the process of transforming a conventional army into a technology-based modern military.

These constraints, however, have not prevented the PLA from pursuing some aspects of the RMA in order to improve China's ability to fight and win a "local war under high-tech conditions." In accordance with this concept, the Chinese RMA emphasizes research and development (R&D) of key technologies, such as advanced laser/particle beam technologies and precision terminal guidance technologies. According to the PLA, a breakthrough in these technologies will enable China to produce laser weapon systems to enhance anti-air weapon capabilities. Precision terminal guidance technology could enhance the lethality and survivability of China's strategic weapons. The Chinese RMA also emphasizes developing electronic IT and integrating it with precision-guided weapons. Information and precision-guidance technologies are considered "force multipliers," an indispensable element of modern warfare under high-tech conditions. IT and artificial intelligence are not only shaping new and revolutionary tactical and operational concepts, but also assisting China in closing the technological gap between China and more militarily advanced Western countries.[4]

The impact of the RMA on China's defense planning in the last decade can be summarized as follows: First, Chinese planners believe that the technology and weapon systems revolution have been instrumental in transforming the PLA from a backward people's war army into a modern military. Second, officers have argued that technology and weapon systems modernization should be selective and emphasize technology that can produce "pockets of excellence," such as electronic and information warfare, precision-guided munitions, and laser weapon systems. Third, planners realize the need to restructure the armed forces, developing new tactics, and doctrine to improve operations under high-tech conditions.

CHALLENGES IN PURSUING AN RMA STRATEGY

The RMA has greatly changed the international strategic environment and many Chinese defense planners have recommended the PLA transform based on RMA concepts. China faces seven challenges in developing a new strategy based on the RMA.[5] The first challenge is conceptual reform. The RMA requires military leaders to adopt new military thinking based upon the information revolution and its implications for the military, and to break away from conventional, mechanical military concepts. The second challenge involves ensuring that IT becomes a leading source of modernization and transformation, based on the recognition that only IT can transform conventional

forces into a modern integrated joint force. A third challenge is to accelerate
IT development to ensure victory in war under high-tech information warfare conditions. The PLA will inevitably confront such an enemy in the
future.

A fourth challenge is to support the ongoing IT-based modernization
efforts. RMA advocates argue that weapon systems modernization efforts
should utilize IT to enhance the firepower, mobility, and defensive capabilities of forces. Current priorities in IT-based weapon systems development
programs include:

1. Development of a military information network, including a command
   and early warning system, and a comprehensive information protection
   system to ensure the reliability and effectiveness of C4ISR systems.
2. Design and development of precision-guided weapons, including missiles
   and long-range precision-guided munitions.
3. Design and development of C4ISR systems, such as reconnaissance,
   communication, and early warning satellites and AWACS airplanes.

To pursue these goals, China is importing advanced technology, reverse
engineering foreign designs, and integrating systems so that a new generation of IT-based weapon systems can be formed into standard packages to
conduct joint operations.

A fifth challenge for China is to improve the PLA's overall military structure
and consolidate military organizations. Some have suggested reducing the over-
sized ground force and shifting resources to the Navy, Air Force, and Second
Artillery. RMA advocates also argue that conventional units should be replaced
with IT-based units. At the command and control level, reforms should ensure
the use of IT to achieve a joint operations capability. The objective of this structural reform is to transform current mechanized combined forces into informatized combined forces in order to improve the PLA's war-fighting capability.

Sixth, an IT-based force structure cannot be realized without reforms in
education and training. Some have suggested consolidating current military
and service academies, and boosting science and technological education to
produce a new generation of IT-oriented officers and servicemen. The PLA
is desperately in need of university graduates trained in advanced science and
technology. Another suggestion is to increase the numbers of PLA officers
with master's degrees and Ph.D.s in science and technology. Educational
reforms propose that PLA officers be trained at civilian higher educational
institutions, including foreign universities. Further, reforms in the recruitment system and the pay structure may attract more talented individuals to
enter or teach at military educational institutions. Training reforms emphasize use of IT-based training facilities, including information networks,
computer-based simulation programs and software, and information-based
training facilities and weapon ranges, for joint operations.

A final challenge is to develop IT-based logistical support systems. In a modern war-fighting environment, "information, protection, and transportation"

must be integrated under one network to provide armed forces in joint operations with effective and timely logistical services and equipment support.

These are the key areas in which China is pursuing the RMA. It is difficult to follow the PLA reform process because little information is available to outsiders. The PLA has debated and developed IT-based operational concepts and theories, but few details are available. China has developed some IT-based weapon systems indigenously while also importing and reverse-engineering advanced technologies and systems. The PLA plans to establish small experimental units of IT-based forces to implement new tactical and operational doctrine, and to acquire experience in an IT environment. The most prominent and visible indicator of China's IT-RMA effort, however, is its high-tech and IT-based weapons development programs. Tracking advanced weapon systems acquisition and development provides a bird's eye view of Chinese RMA endeavors.

China's Development of
Advanced Weapon Systems

As early as the mid-1980s, the PLA's leadership had begun to set parameters to modernize the armed forces in order to fight and win a local war under high-tech conditions and to upgrade the quality of the PLA to the level of advanced countries. Two lines of thinking dominated the debate over modernization. The first argued that priority should be placed on the development and improvement of high-technology elements with strategic importance such as the space industry, laser and particle weapons, automation, biotechnology, information systems, the energy industry, and material sciences. This school of thought, led by top civilian and military scientists, emphasized the PRC's development of advanced technology in the twenty-first century. The other school focused more on military interests and argued that China should give priority to the development of hard-kill and soft-kill information warfare capabilities. Top military planners and theorists in the Central Military Commission (CMC) and National Defense University who consider IT to be the key to upgrading the PLA's C4ISR capability were the chief proponents of this view.

The PRC's leaders do not believe that these two lines of RMA thought contradict one another. In fact, if coordinated and integrated, they could complement each other. In March 1986, as part of "Project 863," the Beijing leadership approved the development of strategic technology. The goal was to upgrade China's high-technology infrastructure and consolidate the existing state-owned military–industrial complex with R&D laboratories in order to enhance China's technological sophistication. In addition, in 1995 as "Project 95" the leadership approved development of an information warfare capability. This program emphasized soft-kill IW capabilities, an enhanced C4ISR network, and the ability to conduct electronic/magnetic warfare, psychological warfare, and cyber war. China's development of a hard-kill IW capability focused on development of "pockets of excellence," such as laser/particle beam weapon systems, anti-satellite weapon systems, cruise missiles, anti-radiation missiles, and laser-guided smart bombs.

The Weapon Systems Revolution: Developing "Pockets of Excellence"

The Chinese leadership has decided to develop "pockets of excellence" or "assassin mace" (*xa xou jian*) weapon systems and technology to deter or defeat a stronger adversary in future war scenarios, particularly with respect to the Taiwan Strait conflict. Open sources indicate that China is committed to developing electronic warfare (EW), information warfare (IW), land-attack cruise missiles (LACM), theater ballistic missiles, military space systems, and even directed energy weapons. Western experts see these developments as part of Beijing's effort to develop an "asymmetrical warfare" capability and pursue the RMA[6] in order to facilitate preemptive strikes. Understanding current and possible future "pockets of excellence" should help in assessing the PLA's challenge to Asian security.

Electronic Warfare

China's EW modernization is based on improving its intercept, direction finding, and jamming capability. In recent years, China has imported state-of-the-art technology from the West and Russia. China is interested in improving imagery reconnaissance and surveillance as well as electronic intelligence (ELINT) collection. It is also interested in developing unmanned aerial vehicles (UAV) to serve as platforms for an improved electronic warfare capability. China's existing satellite earth stations can be modified to interfere with satellite communications. The PLA is developing an electronic countermeasure (ECM) doctrine and has trained in an ECM environment. The PLA also possesses the ability to deploy electromagnetic pulse (EMP) bombs to paralyze the enemy's electronic and information systems.[7]

Information Warfare

China has recognized that information warfare will be a key strategic weapon in future wars, and that their military must develop not only offensive IW capabilities, but also IW defensive measures and countermeasures. Indeed, China has placed greater emphasis on defensive than offensive IW capability, and over the past few years the Communication Command Academy in Wuhan has emerged as one of the PLA's major IW research facilities. The PLA's research in IW is said to be mature, and the Chinese military is currently developing its own IW doctrine.[8] Some reports indicate that the PLA has set up a "net force" (*Wan Jun*) to conduct IW, but there is no evidence to support this claim. Rather, it is believed China is giving priority to developing its defensive IW capability, including computer anti-virus solutions, network security, and advanced data communication technologies. Such efforts would increase the PLA's expertise in defending its own networks against enemy attack and enhance its offensive capabilities in the future.

Cruise Missiles

Open sources indicate that China has a particular interest in producing indigenous land-attack cruise missiles, including programs for airframe design, propulsion systems (such as small turbojet engines and ramjets/scramjets), and guidance technologies (such as GPS for in-flight navigation and terrain contour matching guidance, infrared imaging, or synthetic aperture radar for terminal homing).[9] China reportedly plans to produce at least two types of LACM, Hong Niao-1 and Hong Niao-2, with a probable range of 400–600 km.[10] These weapons are believed to be based upon cruise missile technology and subsystems acquired from abroad, particularly from Russia, and could be deployed sometime in 2005. Meanwhile, China is also improving and modifying LACM technologies for airborne and submersible launch platforms, which could greatly enhance China's long-range strike capability.

There are currently six types of anti-ship cruise missiles (ASCMs) in the PLA's inventory. These include the CSS-N-1, CSS-N-2, CSS-N-4/C801, SS-N-61, C802, C-701, and SS-N-22/ SUNBURN supersonic anti-ship missiles deployed aboard the two Sovremennyy-class-guided missile destroyers (DDGs) purchased from Russia. The indigenously produced ASCMs can be deployed and launched by naval coastal defense missile units, bombers, surface warships, and submarines, providing China with area denial and sea control capabilities.[11]

Space and Theater Missile Development

China has greatly expanded its military use of space in recent years. It is believed that China has the capability to produce and eventually deploy advanced imagery reconnaissance and earth resource systems with military applications. The China–Brazil Earth Resources Satellite (CBERS), launched in October 1999, supports Beijing's efforts to develop improved military reconnaissance satellites.[12] In addition, China has launched at least three low-orbit meteorological satellites (*zi yuan*), two geosynchronous weather satellites (*fong yung*) and two experimental navigation satellites (*bei dou*) in the last few years.[13]

These satellite programs have allowed China to improve its satellite technology for reconnaissance, surveillance, and targeting capabilities. China's exploitation of manned space operations remains a high priority and could contribute to improved military space systems over the next 20 years. China is also seeking to integrate GPS and the Russian Global Navigation Satellite system (GLONASS) guidance technology into fighters, helicopters, and cruise missiles. China can be expected to launch more indigenous navigation satellites in the next few years to improve its long-range strike capabilities.

All ballistic missiles in the inventory of the PLA's 2nd Artillery are capable of carrying both nuclear and conventional warheads. China views its growing conventionally armed Short Range Ballistic Missile (SRBM) force as a potent military and political instrument for dealing with the Taiwan Strait equation, even though doubts remain as to whether or not using missiles to intimidate China's neighbors can advance Beijing's objectives.

China currently possesses two types of solid propellant, road mobile SRBMs: the CSS-6 (M-9) and CSS-7 (M-11). The PLA has at least one regimental-sized M-9 SRBM unit deployed in Fujian province, while an estimated 250–300 M-9 SRBMs are stationed opposite the Taiwan Strait.[14] Improved, longer range variants are available and operational, and satellite-assisted navigation technologies have improved SRBM accuracy since the 1996 exercises in the Taiwan Strait. In an armed conflict with Taiwan, China's SRBMs and LACMs would likely be used as preemptive strike weapons to target air defense installations, airfields, naval bases, C4I nodes, and logistics depots.

Beijing is reportedly developing state-of-the-art anti-missile systems to defend against cruise missiles and Theater Ballistic Missiles (TBMs). To date, China has acquired limited numbers of SA-10B, SA-10C, and SA-15 SAMs from Russia to fill gaps in its air defense structure. China has modified its Hong Qi-7 (HQ-7) series of SAM with an improved capability to counter cruise missiles. The HQ-9 SAM, currently under development, is believed to have been modeled after the U.S. *Patriot*. It is reportedly intended to provide long-range defense against fixed-wing aircraft as well as against TBMs. Though Russia may be involved in developing ABM systems for China, a Chinese ABM capability will not be functional during the next 10–20 years.[15]

Directed Energy Weapons

China is believed to have a highly developed electro-optical industry as well as the ability to build laser weapons, including tactical lasers. China produced the ZM-87 neodymium laser blinder for use primarily against ground targets, but it could also be used against aircraft with improved range and anti-sensor capabilities. China is reportedly investigating the feasibility of ship-borne laser weapons for air defense.[16] Future laser systems will likely emphasize improved target acquisition and pointing and tracking.

The PLA's Changing Force Structure: Rapid Reaction Forces

Ground Forces

Current evidence shows that the PLA has designated seven Group Armies (GA), the 15th Airborne Corps, and two marine brigades as rapid reaction forces (RRFs), also known as "Fist" forces. Of the seven RRF group armies, two are stationed in the Beijing Military Region (MR) (27th GA, 38th GA), one in the Shenyang MR (39th GA), one in the Jinan MR (54th GA), one in the Nanjing MR (1st GA), one in the Lanzhou MR (21st GA), and one in the Chengdu MR (13th GA). The total number of their personnel ranges from 75,000 to 89,000 and their force structure includes tank and mechanized divisions, motorized divisions, infantry divisions, artillery divisions, ballistic missile brigades, anti-aircraft artillery (AAA) and air defense brigades, engineering brigades, helicopter teams, chemical warfare regiments, communication regiments, and electronic warfare units.[17]

RRF GAs contain various combinations of forces, partly due to their mission objectives and partly due to the differing pace of modernization of the various units. For example, the Beijing MR's 38th GA and the Shenyang MR's 39th GA contain more armored and mechanized divisions than other armies, indicating that their mission is the defense of northern China. The 38th GA even contains the latest FM-80 SAM missile brigade and the Z-9 helicopter gunship team (*da dui*), units that are probably allocated to the defense of Beijing. The Jinan MR's 54th GA is a strategic reserve army designed to deploy rapidly to assist other MRs in an emergency. One armored infantry division is stationed near the Beijing Guanzhou and Longhai railway lines and close to the base of 15th Airborne Corps so that it can deploy rapidly to any other part of China, including the coastal area near the Taiwan Strait. The 54th GA also contains an army-level helicopter team, an electronic warfare team, special forces, an artillery brigade, and an anti-aircraft missile artillery brigade. The CMC directly controls the 54th GA, which can be mobilized within six hours and deployed to Beijing or the coastal region near the Taiwan Strait within 15 hours.[18] The Beijing MR's 27th GA is also a strategic reserve unit, but it contains less armored and mechanized divisions and has no helicopter team. As a strategic reserve GA, some units were ordered by the CMC to participate in the 1996 Taiwan Strait exercises. In the case of a Taiwan Strait confrontation, the 1st GA of the Nanjing MR (covering the northern flank of the coastal region), the 42nd GA of the Guanzhou MR (covering southern flank of the coastal region) and the 15th Airborne Corps could form the main attacking forces following missile strikes and air strikes. The strategic reserve of the 27th and 54th GAs could assist frontline operation if necessary. The 12th GA is the main backup for conducting a second wave of attack.

Though RRF GAs are becoming the elite units of the PLA ground force, their main assignment is still the defense of China and land border areas simply because they are heavily dependent on road and rail transportation for rapid deployment. Lack of adequate airlift and sealift capabilities prohibit the deployment of heavily armed RRF GAs beyond China's borders. In the case of a need to confront a capable adversary offshore, it is likely that only those technical and special force units that could be deployed rapidly to the scene by helicopter and airlift would likely be used. Homeland defense rather than force projection will remain the primary mission of the RRF ground force during the next decade or two.

RRF: Airborne Corps

The 15th Airborne Corps is one of the PLAAF's RRFs. It contains three airborne divisions with 30,000–40,000 troops. Each airborne division contains 2–3 regiments and 1 divisional headquarters. The 43rd Airborne Division, designated as a rapid-deployment division and stationed in the Beijing MR, contains one infantry airborne regiment, one armored/motorized infantry regiment, and one artillery regiment. The 44th division is designated as an airborne reserve force and is stationed in the Nanjing MR. The 45th division, also designated as a rapid-deployment division, contains three infantry regiments and one division headquarters and is the only RRF within the Guanzhou MR.[19] The PLAAF currently has 30 Yu-8 and 20 IL-76 transporters that have the capability to conduct long-range airborne drops. With this airlift capacity, the PLA can only deliver two airborne regiments (including one motorized or artillery regiment) to the drop zone at one time, a very limited striking force if not backed by more support.

RRF: Marine Corps

Currently China has two marine brigades attached to the PLA Navy's South Sea Fleet. Their total manpower is approximately 40,000 personnel. Each brigade has one amphibious tank regiment with Type 63 amphibious tanks and Type 77 APCs. Marine brigades are designated as the Navy's RRF. Due to limited amphibious landing capability, however, the PLA Navy can only put up a maximum of 5,000 marines and a limited number of amphibious tanks and APCs in offshore landing operations.[20]

Enhancing Conventional Strike Capability: Air Strike and Sea Control

In addition to the airborne corps, the PLAAF has eight regiments, one battalion, and five airbases that are designated as RRF units and contain 532 fighters, bombers, and transports of various types. The strategy and tactics of the PLAAF to fight a future war under high-technology conditions have been described as "conducting active defense with offensive initiative to impose air blockades, launch air strikes, as well as conduct joint operations with the ground forces and navy."[21] Some PLAAF leaders believe that in order to allow the Air Force to react appropriately to any situation, including gaining air superiority and launching counterattacks against targets inside the enemy's boarders, the PLAAF should rapidly replace aging aircraft with state-of-the-art fighters, attack aircraft, bombers, reconnaissance aircraft, and special purpose planes. Current fighters, bombers, and attack aircraft allocated to the PLAAF RRF are not compatible with the perception of a high-quality air force, and fall short of meeting requirements for air superiority, ground force operations, and even long-range counterattacks. Yet the PLAAF hardly lags behind other services' modernization. In recent years, it has received new fighters, bombers, and air-to-air and air-to-surface missiles, and is about to produce a new indigenously designed fighter. The PLA leadership has allotted funding for the PLAAF's overhaul, and China's air-strike capability could be improved greatly by the year 2010.[22]

The Navy's RRF, aside from the marine corps, is designed to exercise sea control and sea denial within Chinese territorial waters, Exclusive Economic Zones, and peripheral waters in the Western Pacific Ocean. In terms of the PLA Navy's current force structure, China is less likely to build an ocean-going blue-water navy than adequate surface combatants, submersible vessels, and replenishment and support ships to exercise a limited offshore active defense and conduct limited expeditionary operations in defense of territorial claims such as those in the South China Sea. This force posture will also support a sea denial and anti-access strategy in a future Taiwan crisis scenario, and exercise sufficient sea control given the strong coastal missile buildup near Bohai Bay and along the Shangdong and Liaodong peninsulas. To fulfill these strategic objectives, the PLA is upgrading its modern heavy-displacement destroyers and missile frigates, and acquiring new types of diesel submarines, nuclear-powered attack submarines, and naval aviation to ensure effective operations in blockade, sea denial, and sea control.[23]

PROBLEMS IMPLEMENTING THE RMA

China's defense planners are conscious of the importance of the RMA in transforming the PLA into a modern military, but there are many barriers to the implementation of such a revolution. There is no question that China has acquired an array of advanced weapon systems in recent years, yet most were imported from abroad or indigenously designed and produced through reverse engineering. Senior Chinese military officers criticize weapon system development programs that still emphasize "mechanized characteristics." They feel that the programs should adopt a "leap-frog" approach that concentrates on information integration, and that the PLA should seek assistance from the private sector.[24] Reforming the defense industry and the science and technology research institutes faces funding problems and are low priority in terms of developing IT-based weapon systems.[25] Information warfare-related R&D is focused on internet warfare, upgrading the protection of key domestic communications networks, and the development of jamming devices, but progress has been very limited.[26] One significant problem concerns education and training. The PLA lacks IT-trained officers and men, IT-based education and training facilities, and IT-based curriculum in military its institutions. Officers with master's degrees and Ph.D.s make up less than 2 percent of the officer corps, far behind the level required for RMA transformation. Lack of qualified IT-based officers is considered to be the biggest "bottle neck" in RMA development.[27] The advanced nature of military technology and sophisticated skills needed to operate advanced weapon systems places incalculable psychological pressure on officers and men who are inadequately educated and trained. Finally, the latest move in RMA reform which involves cutting half a million military personnel within the next five years may pose an additional political challenge for Jiang Zemin and Hu Jingtao if resistance emerges from below.

WHITHER THE RMA?

China's determination to proceed with some aspects of the RMA is evident in its pursuit of advanced technology and weapon systems. Beijing's capacity to produce an array of ballistic missiles, LACMs, and ASCMs has not only given its military the capability to conduct long-range strikes in future conflict, but also poses a great threat to Taiwan and neighboring territories. Beijing may not possess the ability to project its power against larger Western Pacific states during the next decade or two due to difficulties in assimilating and integrating modern technologies, and producing state-of-the-art naval surface combatants, submersibles, and fighter aircraft.[28] However, China's pursuit of specific technological and weapon system revolutions does go hand in hand with its "active defense" and "near sea defense" strategies and is a source of great concern. Beijing's improvement in developing advanced information technologies and the ability to integrate these technologies to conduct information warfare are of particular concern to China watchers.[29] Moreover, the PLA embarked on a two-phase modernization plan in the late 1990s. The PLA is scaling down into a smaller, higher quality force that exploits RMA concepts to execute joint operations. It is also focusing on "pockets of excellence," or high-technology R&D applications to enhance its ability to conduct information war and long distance precision strike operations.[30] These RMA-like modernization plans will certainly boost China's confidence in fighting and winning a local war under high-tech conditions in the future.

NOTES

1. Research on the RMA-related issues began as early as the beginning of 1987 when the Academic of Military Science (AMS), a top think tank of the Central Military Commission (CMC), was instructed to conduct systematic research on China's defense in the year 2000 and make policy recommendations to the military leadership. The policy recommendation made adoption and integration of IT into weapon system development a priority in China's defense construction in the blue print of defense 2000. See *Future Defense Construction* (*weo lai de guo fang jian she*), edited by the Department of Planning Organization, Academic of Military Science (Beijing: Military Science Press, 1990) pp. 175–180.
2. Liu, Hong-ji, "Comments on New Military Revolution with Chinese Characters," *National Defense University Journal* 104 (February 1998) pp. 67–69.
3. Gao, Chung-xiang, ed., *Theory on New Military Revolution* (Beijing: Military Science Press, Beijing, 1996) pp. 19–33.
4. Ibid., pp. 220–221.
5. A comprehensive and systematic discussion over the RMA and its impact on Chinese military transformation is presented by Lt. General Huang Xin (PLAAF) in his book *Discussion of New Revolution in Military Affairs and Strategic Counter Policies* (*Lun Xin Junshi Geming Yu Zhanlue Duice*). General Huang identified the RMA impact on strategic thinking and war-fighting capabilities as far as the PLA is concerned. He also made suggestions and recommendations for the implementation of RMA concepts to transform PLA, and they are paralleled in many other defense planners. See Huang Xin, *Lun Xin Junshi Geming Yu Zhanlue Duice* (Beijing: Lian Tien publishing firm, February, 2003) pp. 128–171.
6. The PLA's interest in studying the RMA was largely driven by its quest for high-technology advanced weapon systems and the application of high-technology weapon systems in the 1990–91 Gulf War. According to Michael Pillsbury, the PLA studied U.S. and Russian RMA concepts and doctrines since the mid-1990s,

and NDU/AMS researchers have published over 50 RMA-related articles exploring the RMA in the Chinese military context. Some PLA experts argued that the PLA's interest in the RMA is nothing more than an attempt to improve its technology level, which differs greatly from the western understanding of the RMA. However, Pillsbury has managed to sum up the PLA's vision of the RMA, which includes:

(a) to enable the PLA to eliminate the information gap;
(b) to establish a net system to integrate all branches of armed forces;
(c) to be capable of attacking and paralyzing the enemy's C4I;
(d) to ensure attack first;
(e) to employ directed weapon systems;
(f) to develop computer viruses;
(g) to adopt submarine-launched weapon systems;
(h) to adopt anti-satellite weapon systems;
(i) to intercept the enemy's expanding logistics supports;
(j) to deploy special forces in attacks.

See Michael Pillsbury, *China and Revolutionary Military Affairs*, report prepared for the Office for Net Assessment (Washington, D.C.: Department of Defense, 1997) p. 6.

7. Again, many Western PLA experts believe China possesses very limited EW capability either in terms of ECM or in terms of intelligence collection, or even in conducting EW against advanced ECM adversaries. It is witnessed by China's aggression in seeking advanced EW systems from other countries such as Israel and Russia and its collaboration with Russia to produce improved ECM-capable fighter radar and ECM pods. One can only identify China placing priority on an EW upgrade; little evidence supports that China's EW capability has been greatly upgraded in recent years. See Bernard D. Cole and Paul H.B. Godwin, "Advanced Military Technology and the PLA: Priorities and Capabilities for the 21st Century," in Larry M. Wortzel, ed., *The Chinese Armed Forces in the 21st Century* (Carlisle, PA: U.S. Army War College, 1999) pp. 169–205.

8. Chinese IW capability comprises three approaches: first, information acquisition, second, information protection, and third, information attack. It is estimated that China possesses sophisticated information gathering capability through its intelligence gathering networks. China is devoting resources to improve its signal intelligence (SIGINT) capability by building advanced communication satellites and developing mini satellites. Both protective and offensive IW measures have been heavily emphasized in recent years to implement PLA IW doctrine to strike strategic targets to end a conflict quickly by destroying the enemy's ability to wage war. State Science and Technology Commission (COSTIND) oversees the development of IW capability, and it collaborates closely with Central Space Committee (CSC) and China Aerospace Cooperation (CAC) in policy decision-making as well as system integration. See Mark A. Stokes, *China's Strategic Modernization: Implication for the United States* (Carlisle, PA: U.S. Army War College, 1999) chapter 3.

9. Richard A. Bitzinger, "Going Places or Running in Place? China's Efforts to Leverage Advanced Technologies for Military Use," in Susan M. Puska, ed., *People's Liberation Army After Next* (Carlisle, PA: U.S. Army War College, 2000) pp. 11–13.

10. Richard Fisher, "Foreign Arms Acquisition and PLA Modernization: Appendix," in James R. Lilly and David Shambaugh, eds., *China's Military Faces the Future*, (Armork, NY: M.E. Sharpe, 1999) p. 131, also *Flight International* (a weekly aviation magazine) published a photo showing HN-1/2 LACM under test firing in China. See *Flight International* (April 28, 2000) p. 18.

11. Bitzinger, "Going places or Running in Places? China's Efforts to Leverage Advanced Technologies for Military Use," pp. 18–19

12. China's effort in developing improved military reconnaissance satellites is an increasingly major concern of U.S. defense analysts. It is estimated that by the 2005–10 timeframe, China's space-based surveillance systems could have at least four components: (1) synthetic aperture radar (SAR) satellites for all weather, day/night monitoring of military activities; (2) electronic reconnaissance satellites to detect electronic emissions in Western Pacific; (3) mid-high resolution electro-optical satellites for early warning, targeting, and mission planning; and (4) a new generation of high-resolution recoverable satellites for intelligence and analysis. See Mark A. Stokes, "China's Military Space and Conventional Theater Missile Development: Implications for Security in the Taiwan Strait," in Susan M. Puska, ed., *People's Liberation Army After Next* (Carlisle, PA: U.S. Army War College, 2000) pp. 112–113.

13. The ROC armed forces intelligence estimation in China's Space programs speculated China will conduct a series of military satellite launches in 2002–05. Priority will be given to electro-optical satellites for early warning and targeting. See *United Daily* (February 24, 2000) p. 1.

14. The number of M-9 (DF-15) SRBMs deployed in the coastal region of Fujian province is always debatable. Western analysts tend to have a more conservative estimate, while the ROC intelligence community tends to come up with a higher number of SRBMs than Western estimates. An intelligence estimation of the number of M-9 SRBM launchers possessed by the 2nd artillery brigade off Taiwan Strait is between 18 and 20, which means the maximum number of SRBMs fired on Taiwan can be no more than 20 missiles at one time

15. It is understood by this author that China does not have the resources to develop ABM, nor does China show great anxiety over U.S. NMD programs. It is not in China's interest to compete with United States in advanced missile and space programs. China is interested in possessing limited "pockets of excellence" weapon systems, making China a recognized regional power. Also see Mei Lin, "Mission and Primary Battle Technique of Conventional Guided Missile Units of Second Artillery," *Studies on Chinese Communist Monthly*, 35 : 4 (April 2001) pp. 86–98.

16. Though China does have high-power laser R&D programs, it is difficult to say China will deploy space-stationed laser weapon systems in the distant future. See Stokes, "China's Strategic Modernization: Implications for the United States," p. 119.

17. Sources from http://www.cqch.com.cn/zgjs/lj.htm

18. This is the estimation and speculation made by the same website as above. In reality, there is no evidence to support such assessment.

19. There were reports that airborne divisions have been restructured to airborne brigades to enhance its mobility, yet there is no evidence supporting this change. See Milton Wen-chung Liao, "PLAAF Strategy and Weapon Modernizations," in Milton W.C. Liao, ed., *Thesis Collections in China Military Studies* (Taipei: Institute of Chinese Communist Studies, 2001) pp. 374–378.

20. Wen-Chung Zhai, *Taiwan's Survival and Maritime Power Development* (Taipei: Mai Tien Publishing Co, 1999) pp. 167–177.

21. Kenneth Allen, "PLA Air force Organization," paper delivered at the CAPS/RAND PLA conference (June 22–24, 2000) pp. 7–8.

22. Milton W.C. Liao, "Theory and Practice of PLAAF Offensive/Defensive Strategy," *Studies on Chinese Communist Monthly* 35 : 5 (May 2001) pp. 75–77. Again, Milton Liao made very interesting remarks in this recent article. He argued that during late 2000, when PLAAF was asked by the CMC to design plans and the budget for the PLAAF 10th Five Year Plan construction, PLAAF was said to have requested RMB 120 billion (equivalent to U.S.$15 billion at current exchange rate) from 2001 to 2005 to replace aging fighters with advanced jet fighters such as Su-27, Su-30, and J-10, and so on. The number of the first line jet fighters requested by the PLAAF are up to 1200 by 2005. If this estimation is realized, PLAAF's strike capability could be enhanced greatly.

23. Two important trends in the PLA Navy's improved strike capability are missilization of its surface combatants/submarines and the receipt of new air defense/strike aircrafts. The PLA Navy has received three indigenously built diesel submarines: the Song class (Type 039), which is believed to be capable of launching ASCM (probably C-801/802). It is believed Type 039 subs will replace the old Ming-class and Romeo Class submarines before 2010 and become the PLA Navy's backbone task force in sea control and sea denial. Naval Aviation has continuously received new fighter jets such as J-8II, Su-27, Su-30, and FB-7, greatly enhancing its anti-ship capability. See *Yearbook on Chinese Communist Studies 2000* (Taipei: Institute of Chinese Communist Studies, 2000) pp. 5–116.

24. Major General Wang Baocun of the Military Science Academy made such criticism when interviewed by *Liao Wang Xinwen Zhoukan* (Outlook Weekly). See Liaowang Xinwen Zhou Kau (Outlook Weekly) no. 28 (July 14, 2003) p. 21.

25. Taiming Cheung, "Reforming the Chinese Defense Industry From the late 1990s to 2003: How Far, How Fast, and How Successful?" a paper presented at CAPS-RAND International Conference on the PLA Affairs, August 15–17, 2003, Grand Hyatt San Diego, U.S.A., pp. 9–10.

26. US Department of Defense, *2003 Annual Report on the Military Power of the People's Republic of China* (Washington, D.C., July 28, 2003). http://www.defenselink,mil/publs/20030730chinaex pp.31–35.

27. Major General Ku Guisheu of the Chinese Defense University pointed out problems of IT-based education programs and a lack of high-tech. trained officers in an interview with *Liao Wang Xinwen Zhoukan* (Outlook Weekly). See *Liao Wang Xinwen Zhoukan*, op. cit., pp. 23–24.

28. Paul Dibb, "The Revolution in Military Affairs and Asian Security," *Survival* 39 : 4 (Winter 1997–98) p. 110.

29. Lin, Chin-jing, "Military Purposes of PRC's Information Warfare," *Studies on Chinese Communism Monthly* 34 : 11 (November 2000) pp. 111–112.

30. In August 1998, the Chinese Communist Party Politburo approved a national security construction project called "998 National Security System Engineering" in which the PLA was asked to speed up the development of laser/particle beam weapon systems as a deterrence to the U.S. National Missile Defense (NMD) programs. See http://cokin.myrice.com/yaowen/yaowen 200059.htm

7

TAIWAN AND THE REVOLUTION IN MILITARY AFFAIRS

James Mulvenon

INTRODUCTION

Not all Revolutions in Military Affairs (RMA) are created equal. One interesting dimension is the depth of information technology (IT) penetration in a given country. For countries with high levels of IT and a well-educated, information-savvy population, RMAs have the potential to be "deep," forming layers of links with the private sector and feeding off the technological dynamism of industry. For developing countries, however, RMAs can be imported at high cost, but the potential impact is likely to be more "shallow." Using this distinction, the Gulf War might therefore be seen as a conflict between a deep RMA (United States) and a shallow RMA (Iraq). At first glance, the emerging conflict between China and Taiwan appears to also follow this pattern, with Taiwan enjoying a potentially deep RMA and China importing the more shallow variety. While an enormous amount of research has been conducted on the Chinese side of the equation, little if any attention has been given to RMA trends in Taiwan. This chapter seeks to fill this gap, examining both the technological base and military policies of Taiwan to forge an assessment of its military modernization. *The principal finding is that Taiwan is blessed with many technological and economic precursors to a deep RMA, but that largely political and bureaucratic constraints have thus far impeded a full exploitation of this potential capability.*

This chapter is divided into two large sections. The first addresses the social, economic, and technological base in Taiwan to support an RMA. The second outlines the current state of the RMA in Taiwan, beginning with an assessment of the Chinese threat that is driving the process. The section then proceeds with an analysis of overall RMA policy, followed by an examination of Taiwanese programs in a variety of RMA-related areas, including electronic warfare (EW), Command, Control, Communications, Computers, Intelligence, Surveillance, and Reconnaissance Systems (C4ISR), and computer network warfare.

Taiwan and the Information Revolution:
Society, Economy, and Technology

The roots of the RMA in Taiwan are non-military in origin, and instead derive from the dramatic information revolution underway on the island. By almost any metric, Taiwan is one of the most "wired" countries in the world. In terms of demographics, Taiwanese society is a cutting-edge, "early adopter" with a deep penetration of IT, and conscripts from this population bring a high level of comfort with technology to their military service. Taiwan's overall teledensity (telephone subscribers per 100 populations) is 164.78, second only to Luxemborg.[1] ITU figures in 2002 reveal that the per capita subscription to mobile cellular services is now the highest in the world at 106.45 per 100 inhabitants. In comparison with other wired countries in Asia, Singapore's cellular teledensity is 79.14 percent, Japan's is 62.11 percent, and Hong Kong's reached 92.98 percent by the end of 2002. Nearly 64.6 percent of the public switched telephone network subscribers also own cellular phones.[2] As for the Internet, Taiwan's active Internet users in December 2002 numbered 8.59 million. In comparative terms, Taiwan in 2002 ranked eleventh in the world in terms of the number of Internet users, twenty-first in terms of Internet penetration at 38.25 percent, and eighth in the world (third with regard to the Asia-Pacific region, just after Japan and Australia) in terms of Internet hosts, which totaled nearly 1.71 million.[3] In 2002, the island ranked fourth globally in broadband penetration at 9.4 percent, behind South Korea (21.3 percent), Hong Kong (14.6 percent), and Canada (11.5 percent).[4]

Not surprisingly, Taiwan's information-savvy population enjoys one of the world's most advanced information infrastructures, and many features of this infrastructure can be considered dual-use assets for a military pursuing an RMA. Taiwan's civilian telecommunications infrastructure consists of a nation-wide network of fixed telephone lines (coaxial and fiber optic), microwave, wireless (satellite, cellular, paging), and TV and radio broadcast. Taiwan is rapidly developing its telecommunications infrastructure with the goal of becoming an Asia-Pacific telecommunications hub, and the Taiwan military is likely to benefit from any improvements to the commercial architecture.[5]

At an economic and technological level, the Taiwanese industrial and service sectors are world leaders in IT, and provide an outstanding base for the research, development, and production of most RMA-related systems. Yet the nonmilitary RMA environment in Taiwan is not entirely positive. The modernity of Taiwan's telecommunications infrastructure and the ubiquity of IT in Taiwan society represent some of Taiwan's greatest potential vulnerabilities. Like all nations that are heavily reliant on a computer-driven way of life, the island is potentially vulnerable to critical infrastructure attacks, ranging from computer network attack to electronic attack to electromagnetic pulse weapons to fifth column sabotage.[6] Moreover, Taiwan's advanced information industry is a necessary but not sufficient condition for a deep RMA. The island's diplomatic isolation, coupled with a highly conflictual domestic political system, distorts its modernization process, preventing full actualization of the island's potential RMA capability.

THE RMA AND THE TAIWANESE MILITARY

Overall Assessment

The Taiwan defense establishment has publicly expressed its commitment to the implementation of an RMA. The primary driver and motive for the island's RMA is China's military modernization and growing coercive threat. At the same time, Taiwan's RMA process intersects with the ongoing evolution of its military strategy from a defensive to offensive orientation, including elements of EW, C4I and information warfare (IW). Currently, the Republic of China (ROC) military is an early stage of the RMA, reforming its organizational structure and upgrading its core command, control, communications, computers, and intelligence (C4I) structure to facilitate future implementation of a full-spectrum RMA. These reforms are proceeding gradually, aided in part by augmented U.S.–Taiwan military-to-military exchanges, arms sales, and interoperability assessments, which are the primary vehicle for disseminating RMA innovations and thereby influencing the trajectory of Taiwan's RMA. Because the American military and its Taiwan counterpart confront different challenges in terms of missions and required capabilities, however, the Taiwan RMA can be more selective in emulation and adaptation of the U.S. model. While the generalized threat from China, the indigenous IT base, and high levels of IT literacy among the conscript force enable the process, the RMA in Taiwan also confronts specific political, economic, and social obstacles and constraints, including a sluggish economic recovery, high levels of domestic political conflict between the executive and legislative branches over security issues, civil–military tensions, declining defense budgets, bureaucratic resistance to defense reform, and differing perceptions between Taipei and Washington over the imminence of the military threat posed by Beijing.

The Milieu: The China Threat

The RMA in Taiwan, like most military revolutions, is driven largely by outside factors, in this case China's military modernization and increasingly bellicose attitude toward political trends on the island. Democratizing trends begun in Taiwan in the mid- to late-1980s have unleashed an unprecedented pluralism in Taiwan, including discussion of formal independence from the mainland, which is anathema to Beijing. In response to these trends, China in the early 1990s initiated a program of military modernization designed to increase the credibility of its ability to coerce the island to accept reunification. The People's Republic of China (PRC) began aggressively importing advanced weapons systems from Russia to fill certain Taiwan-related niche capabilities and deploying them in the provinces opposite the island.

Taiwan's political evolution and Chinese military developments intersected in 1995 and 1996, when China responded to the visit of Taiwan's president to the United States by engaging in provocative saber-rattling with large-scale exercises and ballistic missile tests.

As a result of the 1995–96 tensions, China's weapons programs now place an increased emphasis on acquiring capabilities designed to strengthen the credibility of Beijing's military options against two identified "centers of gravity" in a Taiwan scenario: the will of the Taiwanese people, which China hopes to shape with stand-off terror weapons like anti-ship cruise missiles, long-range land-attack cruise missiles (LACMs), and short-range ballistic missiles (SRBMs); and U.S. military intervention in a China–Taiwan conflict, which China hopes to shape with long-range, anti-ship cruise missiles, and submarines. In addition to these conventional systems, the Chinese military appears to be actively exploring a variety of RMA-related technologies, including satellite and reconnaissance technology; electronic, computer, microwave, laser, and other radio transmission technology; and electromagnetic weapons. Current modernization programs seek to realize short-term improvements in anti-surface warfare (ASuW) and precision strike, as well as and longer term advances in missile defense, counter-space, and IW writ large. Beijing also is working to address problems associated with integrating advanced weapons systems into their inventory, weaknesses in C4I, deficiencies in training, and new challenges in logistics, so as to improve the People's Liberation Army's (PLA's) overall warfighting capability.[7]

The Evolution of Taiwan's Military Strategy: From Defense to Offense

In the realm of defense policy, Taiwan is pursuing a variety of objectives, keyed to the need to deter the mainland and to reassure the Taiwan public that it is secure from attack. On the most basic level, Taipei desires to possess a credible deterrent or other adequate countermeasures against all likely PRC military threats, through the formulation of an appropriate military doctrine and related operational guidelines for the ROC military as well as the maintenance of a corresponding force structure and C4I/logistics infrastructure.[8] To this end, Taiwan has embarked on an ambitious program of military modernization, including reform of military strategy, organizational structure, force structure, weapons systems, and C4I infrastructure. While the majority of this modernization is not at all revolutionary in intent or scope, elements of the RMA pervade nearly every sector.

In the early 1990s, when the ROC government formally abolished its long-standing emphasis on retaking the Chinese mainland, Taiwan's military doctrine shifted from an emphasis on unified offensive–defensive operations (*gong shou yi ti*) to a purely defensive-oriented concept (*shoushi fangyu*), which excluded provocative or preemptive military actions against the mainland.[9] This purely defensive posture contained two strategic notions: "resolute defense" (*fangwei gushou*) and "effective deterrence" (*youxiao hezu*). The former concept was largely political and defensive, connoting the determination of the Taiwan military to defend all the areas directly under its control without giving up any territory. The latter concept was more active and forward-oriented, underlining the commitment to building and maintaining a military capability sufficient to severely punish any threatening or attacking force and to deny such a force the attainment of its objectives, thereby deterring it from initiating an assault against Taiwan.[10]

To implement this strategy, Taiwan's military forces needed to succeed in carrying out three key missions, listed in general order of priority: (1) air superiority (*zhikong*) for the ROC Air Force; (2) sea denial (*zhihai*) for the ROC Navy; and (3) anti-landing warfare (*fandenglu*) for the ROC Army.[11] Each of these missions was generally viewed by each service as constituting a relatively separate and distinct task. In other words, Taiwan's defense strategy was not focused on joint warfighting. This was reportedly due in part to the small size of the ROC military, the limited expanse of the battlespaces involved, the limited technical capabilities of Taiwan's weapons systems, and the purely defensive nature of the mission given to each service. It also reflected the severe restrictions on operational capabilities imposed by Taiwan's relatively small defense budget, which did not permit even the most basic, individual mission of each service to be fully implemented.[12] More broadly, the separate warfighting missions of each military service reflected the larger "stovepiped" nature of the ROC military structure as a whole.[13]

In recent years, however, military planners and political leaders have placed an increasing emphasis upon the development of a more robust military deterrence, in response to the growing capabilities of the PRC. For most observers, this shift in emphasis implies a focus on the acquisition of a capable air and missile defense system and a significant number of surface and subsurface naval assets, to deal with the threat to Taiwan's security posed by the growing possibility of air or missile displays or attacks, naval harassment or blockades, and amphibious and air-based invasions of territory under ROC control. The Democratic People's Party (DPP) under the leadership of President Chen Shui-bian, however, has also sought to move Taiwan's strategy from its purely defensive focus to incorporate more tactically offensive and counterattacking elements.[14] On June 16, 2000 at the 76th anniversary of the Chinese Military Academy in Gaoxiong, Chen reintroduced the concept "decisive offshore battle" or "decision campaign beyond boundaries" (*jingwai juezhan*), which had first appeared in the DPP White Paper in November 1999. Almost immediately, Chen's doctrine encountered heavy military resistance, particularly among officers who suspected that the strategy reflected an intention to take the fight to the mainland. To counter these criticisms, Taiwan government officials over the course of the latter half of 2000 and the first months of 2001 have repeatedly offered differing definitions and articulations of the strategy, and the operationalization and implementation of the concept still appears ill-formed.

To better understand these new strains of offensive-oriented thinking and their possible policy trajectories, it is useful to examine some of the DPP's cornerstone policy documents. In its "policy manifesto," the DPP asserts that Taiwan "must maintain an efficient yet credible deterrence force to preempt any belligerent action towards Taiwan."[15] The focus on deterrence rather than defense is an important signal of the break with the immediate past, especially the early years of the Lee Teng-hui regime when defense enjoyed the top priority. The DPP instead clearly seeks to move the battle away from the island:

> Our military strategy should be adjusted from passive to aggressive defense. We should abandon the idea of beachhead operation and replace attrition combat with operations to paralyze the enemy. We need to acquire the ability to disable the enemy from starting a war against us, so as to avoid fighting a war on our own soil and avoid endangering the people's lives and property.[16]

Thus, the goals of this strategy are to acquire the capability to "push" Taiwan's defense deep and wide into enemy territory, thus increasing deterrence, paralyzing the enemy's ability to make war on TW, and avoiding losses on the Taiwan side.

Yet the advocates for this position are also adamant that Taiwan will never fire first, despite the attractiveness of preemption for achieving these goals.[17] They point out that Taiwan has never felt the need to preempt historically, though this may have been largely predicated on U.S. intervention. While not firing first, there is consensus that Taiwan must quickly seize the initiative from their attacker, although the line between preemption and immediate counterattack is sometimes blurred. This gray area is highlighted by Alexander Huang, who argues:

> Based on Chen Shui-bian's campaign platform, Taiwan would not engage in arms conflict with the mainland until deterrence fails. In other words, once PRC initiates a war or shows it is preparing to use force, Taiwan will have the rights to conduct attack targets on the mainland.[18]

Since Chen's official announcement of "decisive campaign beyond boundary" in June 2000, some defense advisors within the DPP conducted an assessment on the legality of "anticipatory self-defense" in international law, analyzing the blurred line between first strike and second strike,[19] the feasibility of preemptive air strike against the mainland,[20] and other theoretical studies. If it can be verified that the PRC intends to attack, they reportedly concluded that it is not entirely illegal under international law to strike first.

Assuming that the PRC does strike first, Taiwan's response options are discussed under the single rubric of *fanzhi*, or "counterstrike."[21] The plan is to survive the first strike through hardening and civil defense, and then conduct a counterstrike. However, there is significant debate within Taiwan about the exact purpose of this counterstrike. All interlocutors agreed that it should be designed to qualitatively degrade Chinese capabilities, though there were two views of the end game. A more aggressive group spoke of the counterstrike as the end game itself, presumably a military "victory" with no political negotiation, while another group saw the purpose of counterstrike as primarily improving Taiwan's position at the negotiating table with the Chinese. The counterstrike itself would involve "decisive action" to destroy any enemy force "before it enters our territory," preferably destroying the enemy deep in its own rear base and paralyzing the military targets on its soil. Guided by the aphorism that "offense is the best defense," the DPP Defense White Paper asserts that "every military target and facility of the enemy that pose a threat to us will be on our list of targets of attack, and we will take immediate and effective countermeasures which may include preemptive measures to effectively destroy or disable the enemy's war machine."[22] Specifically, Taiwan's armed forces would seek to use information superiority, air superiority, and long-range precision strike to destroy the enemy's C3I system, logistics capacity, and assembly areas. These capabilities would be achieved through the use of "long-range, precision, guided weapons," and IW. Missiles in particular would "make accurate strikes deep into enemy territory and put the enemy's most threatening, key political and military targets well within range."

Since Chen's elections, these DPP defense policies have been partially implemented, though resistance from the military has been palpable. The first challenge for officials was providing a concrete definition of "decisive offshore battle," which seemed at variance with the MND's stated policy and the articulated doctrines of the ground, air and naval forces. During his brief tenure as Premier, former Defense Minister and Chief of the General Staff (CGS) Tang Fei reportedly had reservations about "decisive offshore battle" term, but agreed that any defensive war should be kept away from the island.[23] Former CGS Tang Yaoming and Defense Minister Wu Shih-wen also made public statements in the summer of 2000 that seemed to move slightly toward the idea of decisive offshore battle, with Tang Yao-ming conceding that the military's strategy need to move "a little offshore."[24] General Tang Yaoming asserted that "decisive offshore battle" is by no means an offensive strategy, since "it does not mean that we will launch an attack on mainland China. We neither have such a capability at the moment, not have we any such development plan. We have no plan to develop surface-to-surface missiles—an effective offensive weapon."[25] Instead, the gist of "decisive offshore battle" is "foiling invading mainland Chinese forces at sea or in the air over the Taiwan Strait by using joint forces of all military branches and avoiding bringing war to Taiwan and its outlying islands."[26] Beginning from the premise that Taiwan is in a "defensive position and has no initiative whereas the enemy has the initiative in war,"[27] Then Minister Wu advocated a similar interpretation of "decisive offshore battle":

Fighting a "decisive offshore battle" means that we should not let the flames of war spread to our island or our land. Offshore therefore refers to the Taiwan Strait. We will not necessarily solve the problem in the Strait to the west or the east of the central line of the Strait. The concept does not suggest an offensive operation, for an offensive operation would run counter to our present strategy of "resolute defense" to a certain extent.[28]

Still, the MND has yet to adopt the term wholesale, preferring "source strike" to "decisive offshore battle." The former term shows up in the Executive Yuan's newly approved mid-term administrative project proposals for 2001–04, in which the MND specifies its future tasks as focusing on developing joint operations and "source strike capabilities."[29] The difference between "source strike" and "decisive offshore battle" may only be semantic, but perhaps it is significant that the Ministry continues to resist adopting the President's specific terminology.

While there does seem to be a growing consensus in favor of moving the battle off the beaches, there is still a great deal of debate over offensive operations on the mainland. The first official mention of the development of offensive systems occurred during former CGS Tang Fei's closed testimony to the Legislative Yuan (LY) on April 15, 1999, though a military spokesman immediately denied the remark.[30] In examining the motivation for a shift to offense, two rough schools of thought can be identified. The first argues that the acquisition of an offensive counterforce capability is necessary to deter China from launching a conventional attack against Taiwan, and if deterrence fails, to significantly degrade China's ability to sustain such an attack against Taiwan. These forces would consist essentially of several hundred SRBMs, air assets, and possibly even an LACM variant of the Hsiung-Feng II capable of striking China's ports, theater C3I nodes, and missile launch sites. The second group argues that Taiwan must focus on acquiring offensive strategic countervalue capabilities to threaten major Chinese cities in Central and Southern China, such as Shanghai, Nanjing, Guangzhou, and Hong Kong. These would consist essentially of a relatively small number of intermediate-range ballistic missiles (IRBMs) or medium-range ballistic missiles (MRBMs) with large conventional or perhaps even nuclear or biological warheads, intended purely as a deterrent against an all-out Chinese assault on Taiwan. After former President Lee Teng-hui put forward his controversial "special-state-to-state" (liang-guolun) formulation, Kuomintang (KMT) party stalwart and KMT, Inc. bagman Liu Taiying threatened that Taiwan would conduct a surprise missile attack on Hong Kong and the open seas off Shanghai if the mainland used force.[31] Both schools recognize that the success of the effort requires significant enhancement in the hardening of the island, including both active defenses and passive defenses. To this end, the MND claims that the majority of Taiwan's military facilities and weapons depots have been moved underground to avoid possible mainland Chinese attack. The military is also using separation and camouflage strategies to hide or disguise its facilities and installations.

Alexander Huang identifies four counterarguments offered by opponents of this new, offense-oriented mindset.[32] First, international reaction will likely be negative. Analysts argue that it is extremely unwise to "talk" too much about Taiwan's offensive options, especially before acquiring such a capability. A clear pronouncement of the intention of taking preemptive actions against the PRC would be "politically devastating to Taiwan, because it will unnecessarily provoke Beijing and antagonize Washington." Second, "without sufficient deterrents in Taiwan's inventory, Chen's concept would naturally lead to a suspicion in the international community that Taiwan may eventually pursue a nuclear option." Third, no one can outline the exit strategy (*zhongzhan zhidao*) for these options. Fourth, the DPP failed to analyze whether Taiwan would be able to acquire the technologies to implement this new strategy, and failed to weigh the domestic and foreign costs of pursuing such a line. An additional criticism of the countervalue strategy would highlight the vulnerability of the Taiwan missile infrastructure, and the difficulty of achieving deterrence with a limited number of airframes.

Taiwan's Military Strategy and RMA Policies

From both official and unofficial statements, the Taiwanese military seems committed to the RMA. As early as 1996, the MND White Paper called for RMA-related modernization efforts in all three services. The Army in particular was directed to strengthen its EW capability and upgrade its C3I system, while the Air Force was focused on photo-reconnaissance, early warning, and EW planes.[33]

Between 1996 and 1999, the emphasis on RMA developments increased significantly, driven in part by the personal advocacy of General Tang Fei and the interest of the opposition DPP in future warfare. In its 1999 Defense White Paper, the DPP declared that "the Taiwan military should actively develop electronic information combat capacity, and amplify the C4ISR system to ensure our information superiority in the Taiwan Strait."[34] Moreover, they declared that the focus of MND modernization should be the Navy and the Air Force, not the traditionally favored ground forces.[35]

The DPP's prescribed military strategy included the principle "Strike Deep and Wide," which incorporated many aspects of U.S. command and control warfare (C2W). According to this strategy, some of which has been formally proposed since the DPP assumed the presidency, the Taiwan military should focus its energies on developing an IW capability, as well as long-range, precision-guided weapons. In peacetime, this information superiority, command of air space over the Taiwan Strait, and long-range, precision-guided strike capability would be used to maintain a strong deterrence force. In wartime, such assets could be used to seize immediate information superiority and air and sea command in the Taiwan Strait to inhibit and destroy the enemy's command and information systems, as well as sea, air, and logistics supply capabilities. In addition, the DPP report called for the development and deployment of long-range early-warning radar and unmanned reconnaissance vehicles, as well as military surveillance satellites to increase Taiwan's early-warning capability.

The 2000 Defense White Paper issued by the Ministry of National Defense (MND) represents the apex of RMA-directed policies thus far. In its opening paragraphs, the report declared:

> The Revolution in Military Affairs (RMA) is a worldwide trend in military development. The Republic of China's Armed Forces should also follow the trend to promote the RMA.[36]

Reflecting a maturing understanding, the report specifically broadened the definition of revolutionary trends to include not only technology, but also advanced concepts, technological know-how, and organizational structures. In particular, the report highlighted the extent to which technical innovation is producing profound effects, particularly in regard to precision strike, dominant maneuver, space operations, and IW. The last sector received special emphasis in the document, which asserted that "the buildup and enhancement of cyber warfare capability for offensive/defensive operations and related defense technologies will safeguard the nation's security."[37] To meet the challenges of communication and information threat from the enemy and to establish this fighting capability in IW, the White Paper's recommendation was fourfold: (1) to build a defense information infrastructure (DII); (2) to develop a C4ISR infrastructure; (3) to augment security measures; and (4) to develop offensive IW capabilities. Each of these efforts is discussed in more detail in later sections.

More recent comments by President Chen Shui-bian and Taiwan's defense ministers confirm the trend toward the development of RMA-related programs in the Taiwan military. At the 75th anniversary ceremony for the Armed Forces University in June 2000, President Chen Shui-bian identified "upgrading early warning and electronic warfare capabilities" as a main direction for future military buildup and combat preparedness training.[38] In July 2000, then Defense Minister Wu highlighted continuing efforts in modernizing the C4I infrastructure, pressing the point that Taiwan need to build "a really integrated telecommunications system for our three services."[39] Expanding on these remarks, Wu in February 2001 asserted that the armed forces would take advantage of military and private sector fiber-optic, wireless, and fixed-line communication systems to create a multi-conduit system that uses national resources effectively, with a particular focus on "information protection."[40] Moving from defense to offense, however, Wu also highlighted the Taiwan military's continuing efforts to operationalize an IW unit capable of offensive operations as a core effort in developing a future RMA military.

Taiwan's Military Organization and the RMA

Advocates of the RMA often assert that the revolution is as much organizational as it is technological, since new strategies and weapons require fundamentally new forms of military structure. To this end, Taiwan's defense policy also includes efforts to streamline, restructure, and strengthen the organization of the ROC military, in order to ensure more effective civilian control over the armed forces, to more effectively integrate defense planning with the larger priorities of the government's national security policies, to eliminate waste, corruption, and inefficiency in military procurement and readiness, and to increase the military's overall combat effectiveness. These goals are to be accomplished largely through the promulgation and implementation of an extensive set of organizational reform laws and military restructuring programs.[41]

The most important of these laws is the landmark January 2000 National Defense Law (*Guofangfa*).[42] Its mandated changes affect nearly every level of the military, including lines of authority, numbers of personnel, strategic planning, and ratios of civilian and military personnel in the MND. The reforms are seen by many as essential to improving the overall capability of the Taiwan military. At the same time, even advocates of the changes are attuned to bureaucratic reality, and are therefore implementing the measures in a phased approach.

One of the most important changes is an augmentation of the role of the MND. The January 2000 National Defense Law eliminates the current direct link that exists, regarding operational matters, between the CGS and the president. In its place, the new National Defense Law combines the military administration and military command systems into one and designates the minister of national defense as responsible for both systems.[43] This change would thus place the military and specifically the CGS *entirely* under the institutional authority of the MND and may thereby increase the ability of the MND to direct important aspects of defense policy. The CGS would serve as both the military staff for the Defense Minister and commander of military operations under the Defense Minister's supervision. Hence, this revision in the National Defense Law would also expose the CGS to greater legislative oversight, as a leading official of the Executive branch solely under the direct authority of the premier. Other proposed changes would reportedly place the service headquarters directly under the command of the MND and also greatly increase the number and functional expertise of MND offices. If enacted into law, these changes, combined with the convergence of military authority systems under the MND, could significantly shift control over basic military decisions from the General Staff Headquarters (GSH) to the MND.

A second major reform involves the streamlining of the existing Taiwan military force structure. While the implementation of this reduction was initiated before Chen's election, the DPP in the past has been a strong advocate of streamlining and will likely aggressively pursue the implementation of the measures. The primary motivation for the DPP's support of reductions is financial, believing that streamlining will free monies for the purchase of high-tech weapons needed for the Air Force and Navy to hold enemy forces away from the island. In the eyes of DPP experts, the first target for downsizing should be the ground forces, which are bloated in size and not central to the new, forward-leaning strategy.

The third set of reforms centers on strategic and policy planning within the military services, the "joint" staff, and the MND. One historical barrier to the modernization of strategic planning has been the lack of true jointness in the Taiwan system. In terms of jointness and long-range planning, the institutions are still pretty weak. There is an office in the MND that combines the role of inspector general and joint doctrine, though it is really more focused on operational matters like common radios and common equipment. The J3 does some planning, but they reportedly do not have real plans. The most important office is the newly merged J5/*liandubu* office in the MND. This new combined office will have many functions, including day-to-day operations, international security affairs, and longer range strategic planning. According to interlocutors, the first step in the latter effort will be strategic force planning, involving input from outside sources. Eventually, the office reportedly wants to outsource this work completely to civilian contractors. In terms of interoperability with the United States during a crisis, there is also a special office under the J3, called the *lecheng jihua*, that is exploring the topic.

In addition to these organizational reforms, a number of RMA-oriented institutions have emerged. The first is the Communications Electronics and Information Bureau (CEIB), which is the coordinating institution within the Taiwan military for C4I, IW, EW, and other related areas. CEIB closely coordinates with the Chungshan Institute of Science and Technology (CSIST), which is the central research and development (R&D) center for the MND. The Chungshan Institute of Science and Technology (CSIST), established in 1968, is the leading institution for the research, development, and design of defense technology in Taiwan. With its headquarters in Lungtan, Taoyuan County, the CSIST has facilities stretching over nearly 6,000 acres scattered throughout Taiwan, employing more than 6,000 scientists and 8,000 technicians. The institute itself is divided into four major research divisions: aeronautics, missiles and rockets, electronics, and chemistry. In addition, CSIST has six centers for systems development, systems maintenance, quality assurance, materials R&D, aeronautic development, and missile manufacturing. CSIST jointly conducts independent R&D of weapon systems with the Aero Industry Development Center, which is now under CSIST supervision; some manufacturing units of the Combined Services Force; academic institutions; and public and civilian industries. To date, a number of weapon systems have been domestically designed, tested, and produced on a mass scale by the CSIST. These include the Kung-feng 6A rocket, the Hsiung-feng I and Hsiung-feng II Cruise Missiles, artillery fire control systems, naval sonar systems, naval EW systems, and the Tzu-chiang trainer aircraft. The CSIST has produced or plans to produce Tien-kung (SKYBOW) I and Tien-kung II SAMs, Tien-chien (SKY SWORD) I and II AAMs, Hsiung-feng III cruise missiles, and Lei-ting (Thunderbolt) 2000 multi-barrel rockets.[44] Any future indigenous development of RMA technologies, or the integration of imported systems, will be carried out under CSIST's direction.

Military Modernization and the RMA

To successfully implement its new, more offensive-oriented military strategy, Taiwan must augment its limited indigenous military systems by obtaining critical weapons, support infrastructure, and military technology and training from the outside. For more than a decade, Taiwan's military modernization effort has focused on acquiring modern weapons systems and associated equipment to deter—and, if necessary—defeat Chinese aggression.[45] Billions of dollars have been spent on domestic programs like the Indigenous Defense Fighter (IDF) and the Tien Kung air defense system, as well as on foreign purchases like the U.S.-made F-16 fighter and the French-built Lafayette-class frigate. Many of these newer systems are in the process of being assimilated into the active inventory.

At the same time, however, Taiwan has actively pursued R&D of advanced RMA-related technologies. According to the 2000 Defense White Paper, Taiwan's military R&D is meant to focus on electronic battle, air and sea superiority, and anti-landing capability. Three key areas are EW, C4I modernization, and IW.

Electronic Warfare

The first RMA area in which the ROC military expressed an interest was EW. In the mid-1990s, military programs centered on defensive uses of EW, particularly in air, sea, and ground operations.[46] Interest in EW has increased in recent years, thanks to greater PRC interest in acquiring EW equipment, advanced radars, airborne early warning aircraft, maritime surveillance radars, and anti-radiation missiles. In May 1999, Defense Minister Tang Fei testified that mainland China would likely attain an EW supremacy against Taiwan by 2010.[47] At the time, he noted that Taiwan had established a military task force to ensure the nation is prepared for EW. The task force is conducting relevant research and working out preventive measures against an enemy's electronic attack. Soon after, General Lin Chin-ching of the CEIB confirmed Taiwan's vulnerability:

> As to the overall threat, Taiwan's electronic warfare capability remains inadequate. The ROC military has established some electronic warfare capability, including information reconnaissance, defense and electronic strike, with aircraft- and ship-based self-defense electronic warfare capability being its main preparations. In addition, since it is lacking in all other areas, such as electronic parameter intelligence data analysis capability, battlefield electromagnetic spectrum management, anti-satellite reconnaissance, electromagnetic pulse defense, and jamming of enemy precision-guided attack weapons capability, it has included all of these in its planning preparations.[48]

According to the 2000 Defense White Paper, current efforts are devoted to improving communication and electronic countermeasures, developing opto-electronic countermeasures and early-warning capability, and assisting the Armed Forces in the installation of training and testing of equipment.

C4I Modernization

Taiwan's military C4I system reportedly consists of a nationwide network of fixed telephone lines (coaxial and fiber optic) and microwave, as well as satellite, troposcatter and HF/VHF radio. The Taiwan military is currently involved in an active program of C4I modernization. General Tang Fei, former CGS and Defense Minister, is widely credited with emphasizing the

importance of this sector, which previously had not enjoyed the same level
of priority as higher profile political programs involving U.S. arms sales.
Perhaps the most important period in this process was the U.S.–Taiwan military exchanges on battle management and C4I in 2000, which identified
key problems in the system and put forward a set of operational recommendations. Former Defense Minister Wu Shih-wen and CGS Tang Yao-ming
publicly endorsed these recommendations.

Communications. In January 2000, the Taiwan military entered a new
period of military communications when it completed a broadband fiber
optic cable communication network connecting the Hengshan Command
Center in Taipei with frontline combat units.[49] Vice CGS General Miao
Yung-ching presided over the inauguration of the Armed Forces Fiber-optics
Communications System in May 2000, saying that the system will give the
armed forces the ability to "hear, see, and command."[50] The newly completed fiber-optics communication system will allow the defense ministry to
acquire near-instantaneous audio–video intelligence from the frontlines and
conduct operational meetings with its field commanders through videocon-
ferencing. Miao noted that as military components continue to be consolidated after the streamlining of Taiwan's new-generation armed forces,
systems that provide enormous firepower need a precise intelligence and
information gathering system together with a communications system that
can deliver such intelligence and relay operational orders. He said that intelligence departments in the future will no longer have to use "special couri-
ers" because the fiber-optics network will be able to transmit broadband
video, still pictures, and digital data, as well as traditional voice communications from the frontlines. Planning and designing of the system began in
1990, while installation of the fiber-optic cables and equipment, testing, and
transfer to the new system began in June 1997, Miao added.

The most recent C4ISR development has been the long-awaited awarding
of the contracts to upgrade the *Po Sheng* C4ISR system.[51] In September 2003,
the Taiwan government awarded Lockheed Martin the first in a proposed series
of military communications contracts, aimed at providing an umbrella communications system for the Navy, Army, and Air Force. The first contract is initially
worth US$27.6 million, but contains options that could total US$2.15 billion.
Jane's Defense Weekly describes the contract as a "significant step," which could
shape Taiwan's future for many years to come. Under a subcontract, U.S.-based
Gibbs and Cox Inc. will be responsible for designing electronic systems for
Taiwan's Cheng Kung-class guided missile frigates. Space and Naval Warfare
Systems Center (SPAWAR) is responsible for overseeing the program for the
U.S. government, which still hopes that Taiwan will eventually purchase the
original architecture, which was priced at over $3 billion.

Reconnaissance Capabilities. During the 1980s, Taiwan's reconnaissance
capability and 1970s vintage photographic technology was adequate for the
limited capabilities and low threat posture of the PLAAF. Taiwan's airborne

reconnaissance capability, however, began to decline precipitously in the 1990s. Last year, the TAF retired the last of its RF-104G tactical reconnaissance aircraft and replaced them with reconnaissance-configured RF-5E aircraft. Taipei continues to seek a new imaging system capable of exploiting targets at greater distances from the coast, but without exposing its reconnaissance flights to China's increasingly more sophisticated air defenses. Taiwan conducts technical and human intelligence operations against China and purchases French SPOT, U.S. LANDSAT, and possibly U.S. IKONOS II commercial imagery for exploitation.[52] Finally, research is currently being conducted on tactical unmanned aerial vehicles (UAVs), which can be used for day-and-night reconnaissance photography, real-time information transmission, automatic pilot control, and global positioning navigation.[53]

Long-Range Early-Warning Radars. In the late 1990s and 2000, early-warning radars appeared on Taiwan's procurement shopping list. Taipei's official policy views on the matter were revealed in March 1999, when Air Force Deputy Chief of Staff Wang Chih-ke confirmed that purchasing an early-warning radar was "the policy of the Defense Ministry."[54] The main rationale for the long-range early warning radar centers on the Chinese missile threat and Taiwan's hope to deploy theater missile defenses. The Taiwan Ministry of Defense (TMD) argument for the radars has both political and military dimensions. On the political front, additional early-warning of Chinese missile attack is seen as a civil defense measure and a boost to public morale, particularly if it improves the population's chances of survival. On a military level, early warning radar is perceived by advocates in Taiwan and Washington as aiding peacetime monitoring of Chinese air deployments, missile engagement, and dispersal of forces, as well as leadership survival. The radars do offer some ability to track aircraft, but the high rates of commercial air traffic in the area and the curvature of the Earth are constraints.

One obstacle in the procurement process was choosing the appropriate radar. Early on, there was discussion about selling PAVE PAWS to Taiwan, but there was general agreement that such a powerful radar far exceeded the needs of Taiwan, especially given its lack of strategic depth. Instead, Washington conducted studies to determine the island's actual requirements, in order to construct a custom-configured set of radar systems to meet those specifications. Interlocutors in Taiwan describe a system of strategic and tactical radars, with two giant phased-array radars at the north and south ends of the island connected by a set of smaller radars at the sites of missile batteries and other TMD infrastructure. By one estimate, the new radars alone, including perhaps two strategic early-warning radars and a network of tactical radars, could cost NT$26 billion (US$785.7 million).[55] There are reports that this plan has already been approved, though there is no evidence that Taiwan has allocated money to purchase the systems. Given the political support for the transfer, however, it seems certain that the radars will be acquired and deployed as soon as possible.

Information Warfare

Perhaps the most important future RMA capability for the Taiwan military is IW, particularly computer network operations. Taiwan military and civilian leaders since the late 1990s have increasingly identified IW as a core interest. During his presidential campaign in 1999, for example, Chen Shui-bian argued that Taiwan should build "information warfare capabilities."[56] In November 1999, then Minister of National Defense Tang Fei submitted a National Defense Policy Report to the LY, emphasizing the central importance of IW as a new type of war and indicating his intentions to give top priority to IW-related preparations. In March 2000, the Taiwan military's head IW officer Lin Chin-ching publicly declared that "building up information warfare attack and defense capability is obviously the top-priority mission of our armed forces preparation."[57] Answering inquiries from legislators in November 2000, Lin continued this line of argument by asserting that IW would "become a key factor in the balance of power in the Taiwan Strait in the future," adding that then CGS Tang Yao-ming wanted the military's future arms build-up to focus on the creation of an IW force.[58] Six months into his term, President Chen Shui-bian in December 2000 urged the military to further upgrade its IW capability and develop an integrated combined services command and control system.[59]

The main driver of Taiwan's interest in IW is the perceived growing threat posed by China's interest in IW. Advocates of IW in Taiwan note the dangerous IW asymmetry across the Strait, since Taiwan's economy, government, and military are highly dependent on computers and could be vulnerable to such high-tech weapons. These attacks are potentially highly destabilizing to Taiwan, both for physical and psychological reasons. On the physical side, some believe that the PRC could use electronic and computer technologies to destroy or disrupt critical civilian and military infrastructure, including information and communications systems and military command structures, without much expense and loss of life.[60] But the psychological impact may even be more significant, as Chinese IW might bring about social confusion, undermine public morale, spread disinformation, paralyze the financial market, and thereby aid the PRC in successfully coercing the Taipei government to enter negotiations for reunification.

In the summer of 1999, these fears of Chinese IW attack appeared to be confirmed by substantial hacking from the mainland. The impetus for the attacks was then Taiwan President Lee Teng-hui's radio interview to a German news organization, Deutsche Welle in June 1999. In the course of his remarks, he put forward a controversial new formulation for China–Taiwan relations, which he dubbed "special state-to-state relations" (*Liangguolun*). Given the importance of subtle changes of language in the highly charged sovereignty conflict across the Taiwan Strait, Beijing's reaction to this new description was predictably intense. For weeks, both governments exchanged charges and countercharges, threats and counter-threats. The populations at large continued this ferocious debate on the Internet, creating nationalist web pages and exchanging epithets in Chinese-language chat rooms. It was not long before this war of words involved the hacker communities on both sides, and by early August a full-scale hacker war had erupted. The first salvo was actually a piece of psychological

Table 7.1   Known hacked sites (partial list)

Country	Hacked sites
China	Securities and Regulation Commission Science and Technology Commission Ministry of Railways Shaanxi Science and Technology Network Shanghai Huwan Education Bureau
Taiwan	Control Yuan Investigation Bureau National Assembly National Taiwan University Library Industrial Technology Research Institute Administration Government Information Office (Executive Yuan) Council of Labor Affairs Pingtung County Government National Information Infrastructure (NII) Project National Center for Research on Earthquake Engineering National Laboratory Animal Breeding and Research Center National Institute of Preventive Medicine Data Communication Business Group of Chunghwa Telecom Third and Eighth River Basin Management Bureau

warfare that spilled over into the world of reality. On August 6, a Chinese-language website registered in the United States but owned by a Chinese company posted a false news report that a Taiwanese F-5E fighter aircraft had been downed by a Chinese Su-27.[61] The report sent the Taiwan stock market into a downward spiral, dropping nearly 2 percent in a single day.

This event was quickly followed by a rapid series of hacks of government and commercial Internet sites by hackers on both sides of the Strait. Some hacks were the work of individuals, while others represented the handiwork of loose groups of hackers with prosaic names like the "Alliance of Red Chinese Hackers" and the "Chinese Hackers Emergency Conference Center."[62] A partial list of these hacks is compiled in Table 7.1.

On August 9, a person claiming to be from the mainland and reportedly operating from a site in Hunan Province hacked into the website of the Taiwanese Control Yuan, the government's watchdog agency, and posted pro-China messages meant to refute President Lee's "special state-to-state" formulation.[63] One message inserted into the page read "Only one China exists and only one China is needed." Another message asserted that "The Taiwan government headed by Lee Teng-hui cannot deny it."

Taiwanese hackers retaliated by inserting pictures of Taiwan's flag, sound files that played the Taiwanese national anthem, and pictures of Taiwan's presidential candidates on mainland Chinese websites.[64] They also posted statements on the websites: "Reconquer, reconquer, reconquer the mainland," "Counter the Chinese Communists," "Taiwan does not belong to China," and "Seriously, Taiwan is also better."[65]

As time went on, the attacks became more serious. On August 10 and 12, a mainland hacker twice broke into the website of the Taiwanese National

Assembly, paralyzing its mainframe computer.[66] During the first attack, only files were replaced and a flag graphic was placed on the front page, but the second attack reportedly included the introduction of viruses into the system, thereby damaging both hardware and software.[67] Throughout the remainder of the month of August, both sides kept up the assault. At the height of the crisis, CCTV, China's main television station and reportedly a high priority target for Taiwanese hackers, claimed that its servers were being attacked once every three minutes. As August became September, however, the furore began to die down, and the hacker attacks dropped off correspondingly.

In retrospect, the hacker "war" between China and Taiwan involved a significant amount of activity on both sides, though none of it can currently be confirmed as state-directed computer network attacks. The Taiwan National Security Bureau, which is responsible for Internet security on Taiwan, reported that Taiwan suffered 72,000 "attacks" during the skirmish, of those only 165 penetrations were successful.[68] The 72,000 figure must be treated with some caution, however, as it likely represents 72,000 "scans" of Taiwanese systems, hundreds or thousands of which could be automated by a small group of people. There are no comparable figures for the Chinese side, though at least five penetrations could be confirmed. The vast majority of the attacks appear to be a form of "web vandalism," with young people defacing lightly protected external web servers of government offices. It cannot be discounted, however, that the military or security services on both sides of the Strait took advantage of this chaotic environment to conduct more sophisticated and malevolent computer network exploit activities.

In the aftermath of this incident, official estimates from Taiwan's MND assert that China might be able to achieve this capability by 2005.[69] Pentagon estimates also suggest that China has yet to acquire an operational capability:

> Although the PRC has achieved certain results in information warfare tactics, their basic capability and technology in information science and technology is still at the elementary research and development stage. The major reason is that its domestic information industry is still mainly in re-processing manufacture and there is no real research and development capability to be mentioned.[70]

Analysts in both Taipei and Washington speculate that China's future cyber arsenal might include computer viruses and possibly even electromagnetic pulses (EMP) weapons.

Ironically, Taiwan appears to possess the greater potential to develop an operation IW capability, since its IT sector is significantly more advanced than its counterpart on the mainland. Taiwan's possible advantages in this area have not gone unnoticed by outside observers, including the Pentagon:

> IO may be an attractive—but untested tool—in multiplying the effectiveness of Taiwan's military forces. As one of the world's largest producers of computer components, Taiwan has all of the basic capabilities needed to carry out offensive IO-related activities, particularly computer network attacks and the introduction

of malicious code. As Taiwan increases its role in the manufacture of new computer warfighting systems, Taipei's capability to exploit its position for IO activities can be expected to increase substantially.[71]

These advantages are also recognized by IW advocates in the Taiwan military, who point to potential spin-on benefits of IT development on the island:

> In the ROC, the accumulated information science and technology capability in the civilian sector (especially the computer virus attack and defense technology) is very outstanding. Therefore, the ROC military can utilize all the avenues to join forces with industry, government, academia, and research institutes to develop the key technology and assist defense organizations rapidly for establishing information warfare attack and defense capability.[72]

General Lin Chin-ching also highlights Taiwan's demographic advantages, arguing that "Taiwan's information warfare advantage, which cannot be matched by the mainland, is that all of our citizens have a very high level of universal education, with a solid communications infrastructure, and our related research on electronic anti-virus software and Internet defense products all being up to world-class level."

One area of proven Taiwanese IW capability, though not explicitly of military origin, is in the field of computer viruses. Among international virus experts, Taiwan is judged to be a leading laboratory for new strains. For example, a virus known as "Bloody" or "6/4," designed to protest the Tiananmen Square crackdown, was first discovered in Taiwan in 1990. In 1992, personnel from The Hague—with support from INTERPOL—investigated the dissemination of the "Michelangelo" virus by a Taiwan firm. In 1996, Taiwan virus writers developed and distributed a computer virus protesting Japanese claims to the Diaoyutai Islands. The following year, opponents of the Taiwan government developed a widely circulated Word-macro virus known as "Con-Air," which protested social problems on the island. Most of the virus of Taiwanese origin are the CIH virus, whose name is derived from the initials of its Taiwanese creator, Chen Ing-hau.[73] The CIH virus manipulates executable programs in Windows 95 and Windows 98, destroying floppy drives and hard disks.[74] The virus reportedly afflicted 360,000 Chinese computers on April 26, 1999, resulting in more than RMB1 billion in damages.[75] The date of the attack was the anniversary of the Chernobyl reactor disaster in Russia, leading some to relabel the virus as the "Chernobyl" virus. While Taiwan is known for creating viruses, it is also well known for the efforts by researchers and corporations to combat computer viruses. Trend Micro—formerly known as Trend Micro Devices—is an industry leader in anti-virus software and, to a lesser extent, other network security products. Trend Micro was the first company to develop a response to the "Michelangelo" virus; it currently dominates the anti-virus software market in Japan. Trend Micro also has led in the area of virus recognition technology. Taiwan's Academia Sinica also has made impacts in the area of anti-virus software development.[76]

While Taiwan possesses considerable advantages in the IW area, it is also necessary to point out that Taiwan does suffer from some fiscal, bureaucratic, and technical constraints. On a fiscal level, Taiwan's defense budgets have declined in recent years, in sharp contrast to the double-digit growth of the mainland's official defense expenditure. This financial pressure is exaserbated by procurement of large numbers of second-generation fighter aircraft and other expected big-ticket items, such as ships, subs, and theater missile defense systems. In order to support the costs of these purchases, as well as their maintenance expenses, advocates for IW complain that there is very little defense budget left for IW equipment.[77] The second constraint is bureaucratic opposition to IW. This factor is closely linked to the first, since the various service branches view interest in IW as coming at the expense of their conventional procurement programs. Third, Taiwanese IW officials complain of export restrictions on U.S. key technology, which in turn impedes the development of high-technology equipment for IW.

Despite these obstacles, however, IW in Taiwan appears to be gaining ground, both in terms of strategy and operational capability. At a strategic level, the notion that IW will be the trump card in the Taiwan Strait defense operation in the future appears to be approaching military-wide consensus.[78] Even opponents concede that IW might provide Taiwan with more warning time of any irregular military activities and provide it with force multipliers to counterbalance the PRC's EW against Taiwan. Moreover, the civilian leadership has become one of the most formidable proponents of IW in Taiwan. During the 2000 presidential election, Chen Shui-bian offered a defense concept known as "preemptive defense," defined as maintaining strong deterrence posture during peacetime through the development of IW and long-range precision-strike capabilities. During wartime, however, Chen asserted that Taiwan should apply preemptive measures, including to IW, to suppress and destroy enemy's C4I system and its warfighting and logistic capabilities.[79] With his victory over Lian Zhan and James Song in March 2000, it was widely expected that IW would assume higher priority within the MND.

Yet the ministry had already begun to organize the implementation of an IW capability even before Chen's election. In April 1999, advisory committee on IW strategic policy was created under the MND, as well as an advisory body consisting of experts from the private sector. The existence of this IW research and training task force was revealed in May 1999 testimony by General Tang Fei to the LY.[80] Further discussion of this organization occured at a hearing on protection of Taiwan's computer systems from mainland Chinese intrusions, chaired by DPP legislators Lee Wen-chung and Tsai Ming-hsien. At the hearing, General Lin Chin-ching, director of CEIB offered his opinion that "we [Taiwan] are able to defend ourselves in an information war, but we will not initiate an offensive." The proposed committee was organized to invite experts and party representatives to study its comprehensive strategy to combat IW. To this end, the MND on September 14, 1999 launched the first of nine seminars "to beef up the military's ability regarding electronic warfare and to cope with the Chinese Communist threat."[81] The seminar was entitled the "2000 Telecommunications Information Security Lecture." It focused on communications security and computer virus protection, and aimed to "show the ministry's determination to ensure information security." The year 1999 also witnessed significant technical developments in the computer network defense area. According to the 2000 Defense White Paper, the ROC military deployed gateways, secure e-mail, and firewall systems, as well as a system of internal certification management and red teams.[82]

In late 1999, budget support for IW was a critical concern of the military leadership. In November, General Tang noted that should the Legislature approve an increased budget, priority would go toward adequate defenses against IW.[83] To bolster the case for IW, the Taiwan MND laid down short-, intermediate-, and long-term targets.[84] The short-term target was focused on enhancing security of the information system and building a security mechanism. By integrating the various systems, the intermediate-term target was focused on developing capability to develop software and hardware for EW, improving frequency spectrum management, and enhancing electronic reconnaissance, electronic attack, electronic defense, and other EW capability. The long-term target was to continue to improve tactical and technological R&D, and to build an automated and digitized technological electromagnetic warfare. In addition, the military's network and information systems were to be pushed underground with greater shielding, so as to protect them from damage caused by EMP blasts.

Chen Shui-bian's election as president heralded a new era in Taiwan's military IW program. Military leaders now highlight that the next five years will be the critical period in the PRN–ROC IW competition.[85] To achieve eventual dominance, General Lin Chin-ching of the CEIB in March 2000 outlined a seven-point plan:

(1) Network Security.
(2) Establishment of IW Attack Action.
(3) Joint C4ISR.
(4) Transformation of Defense Management Information Systems.
(5) Construction of a High-Quality Communication and Information Environment.
(6) Public–Private Partnership.
(7) Critical Infrastructure Protection.

Progress in the implementation of this system could be seen in the August 2000 exercise known as "Hankuang #16," in which Red and Blue used computer viruses to attack each other's computer systems. At the time of the exercise, the Taiwan military reportedly had gathered 2000 computer viruses.[86]

On January 2, 2001, the Taiwan military formally inaugurated its first IW force.[87] Then Minister of National Defense Wu Shih-wen had revealed the IW force plan to legislators on November 22, 2000 during the recess of a defense budget screening session held by the legislature's Defense Committee.[88] The unit, which is composed of almost one battalion of specialized troops, is independent of any service and is directly controlled by the office of the CGS. The command is in charge of researching and developing both defensive and offensive IW techniques, though the force was established mainly to cope with potential threats from China in the field, said Major General Chen Wen-chien, deputy director of the communication electronics and information bureau, under the defense ministry:

> We started planning for the new unit in 1998 as we saw the vast efforts made by the Chinese military to upgrade its information attack capabilities. It is to operate under the communication and information command of the Ministry of National Defense. The unit exists for two main purposes. The first is to maintain the security of computer networks in use in the armed forces. The second is to contribute to the overall information security of the country using its specialized knowledge of technology. The unit will become more consolidated after recruiting additional specialized personnel from different branches of the armed forces. For the moment, the unit is staffed by personnel with the Ministry of National Defense's recently-decommissioned "unified communication command," the predecessor of the new Communications, Electronics, and Information Bureau. We did not widen our selection of personnel mainly because of restrictions brought by the ongoing Chingshih personnel streamlining project. But the current staff of the unit should be capable of handling their new tasks since what they are doing is almost the same as what they did before. Future personnel will be drawn from across all services.[89]

According to Chen, teams charged with maintaining network safety and pinpointing computer viruses within the military network will be set up in various regions across Taiwan in the future. While these countermeasures are primarily for defensive purposes, the head of this new command was clear about its potential for future offensive operations: "Should the People's Liberation Army launch an information war against Taiwan, the military, armed with 1000 computer viruses, would be able to fight back."[90]

CONCLUSIONS

Driven by the threat from mainland China to its vulnerable information infrastructure, Taiwan has begun to implement an RMA, emphasizing modernization of technical systems, organization, and doctrine to meet new military challenges. Various RMA sectors, including EW, C4ISR, and IW, have received significant emphasis. Specialized units have been established, and comprehensive plans for integration of technology have been put forward. These efforts are bolstered by the island's advanced IT sector, and the demographic benefits of an information-savvy population. Yet the RMA in the Taiwan military continues to suffer from important constraints, ranging from the usual atavistic opposition from status quo-oriented bureaucracies to the *sui generis* consequences of Taiwan's diplomatic isolation. More recent challenges include Taiwan's stalled defense reforms, a steady decline in defense expenditures, a sluggish economic recovery, and the increasingly confrontational role of the LY and opposition politics in defense policy issues.

Even more than the China threat, however, the most important influence on the trajectory and character of the Taiwanese IT-RMA continues to be the U.S.–Taiwan military relationship, which has increased in scope and scale since 1997.[91] Beginning with assessments of Taiwan's C4ISR architecture and upgrade recommendations, the current interactions are focused on implementation of the new *Po Sheng* program contract. In the past two years, however, elements of tension and frustration have arisen between Taiwan and the United States, even while the two countries move closer together in formal ways. A generally more sanguine assessment of the imminence of the China military threat in Taipei has resulted in uneven implementation of defense reforms, declining defense budgets, and reluctance to purchase most of the military equipment approved for sale by Washington. Perceived pressure from the U.S. side to spend more money on arms, including the C4ISR modernization upgrade, has generated defensive responses from senior military and civilian leaders in Taipei, as have stated concerns about the possible negative implications of the transfer of advanced U.S. military ITs into a hostile counterintelligence environment. Even after approval of the initial *Po Sheng* contract, calls continue for Taiwan to purchase the remaining $2 billion of the proposed architecture while domestic legislators insist that Chungshan Institute and other indigenous providers be given a chance to build the additional pieces. This politicization of the IT-RMA process trends can only serve to distract both sides from effective deployment of the systems, and mitigation of these conflicts should be a top priority for both governments, particularly as Chinese military modernization accelerates and the gap between Chinese and Taiwanese capabilities grows dangerously wide.

NOTES

1. Data from International Telecommunications Union, 2002.
2. Ibid.
3. Ibid.
4. Ibid.
5. Office of the Secretary of Defense, *The Security Situation in the Taiwan Strait*, Report to Congress Pursuant to the FY1999 Appropriations Bill, February 26, 1999.
6. 2000 White Paper.
7. Office of the Secretary of Defense, *The Security Situation in the Taiwan Strait*, Report to Congress Pursuant to the FY1999 Appropriations Bill, February 26, 1999.
8. Much of the following description of Taiwan's defense doctrine and related military policies is excerpted from Michael D. Swaine, *Taiwan's National Security, Defense Policy, and Weapons Procurement Processes* (Santa Monica, CL: RAND, MR-1128-OSD, 1999) pp. 51–61.
9. Alexander Chieh-cheng Huang, "Taiwan's View of Military Balance and the Challenge It Presents," in James R. Lilley and Chuck Downs, eds., *Crisis in the*

Taiwan Strait (Washington, D.C.: National Defense University Press, September 1997) pp. 282–283.

10. Ibid., pp. 284–285.

11. The first two missions reportedly enjoy the highest priority, given the importance of air and sea denial capabilities to preventing air or missile attacks, blockades, and invasions and the fact that Beijing is currently stressing the improvement of its air and naval power projection capabilities.

12. Taiwan's defense budget fluctuates between $8 and 10 billion, while the PRC defense budget is generally estimated by most well-informed analysts as somewhere in the range of $30–35 billion. Moreover, due to the increasing cost of social welfare programs and infrastructure investment, the share of Taiwan's defense budget as a percentage of both total government expenditures and GDP has fallen in recent years. And much of Taiwan's defense budget is taken up by huge personnel costs, which greatly exceed both operational costs and military purchases. In the FY99 defense budget, these three categories of expenditure respectively accounted for 50.5%, 19.09%, and 30.86%. Moreover, arms acquisitions represent only a very small portion of overall military purchases. See Ding and Huang paper, pp. 2–3.

13. In recent years, however, a greater emphasis has been placed on developing joint-operations capabilities. Specifically, efforts to develop joint operations have made some significant headway in the areas of C3I and EW/reconnaissance, where jointness is becoming increasingly necessary. For further details, see Swaine, 1999.

14. This section benefited heavily from Alexander Huang's excellent essay "Homeland Defense with Taiwanese Characteristics: On President Chen Shui-bian's New Defense Concept," presented at the 11th Annual PLA Conference, U.S. Army War College, Carlisle Barracks, PA, December 1–3, 2000.

15. "DPP Year 2000 Policy Manifesto Abstract," November 24, 1999.

16. Democratic Progressive Party Policy Committee, "White Paper on Defense," November 23, 1999.

17. In November 2000, Defense Minister Wu Shih-wen asserted "We would never be the first to fire." See "Taiwan Defense Budget Drops to New Low," *Central News Agency*, November 20, 2000.

18. Alexander Huang, "Homeland Defense with Taiwanese Characteristics: On President Chen Shui-bian's New Defense Concept," presented at the 11th Annual PLA Conference, U.S. Army War College, Carlisle Barracks, PA, December 1–3, 2000.

19. Su Tzu-yun, "The Feasibility of Making First Strike and Taiwan's Rights of Anticipatory Self-Defense," conference paper for Air War College Conference on *Air Force 2011*, November 4, 2000, pp. 3–17.

20. Chang Kuo-cheng, "On the Execution of Air Operations Beyond Boundary," paper presented at the Air War College Conference on *Air Force 2011*, November 4, 2000. Mr. Chang is deputy director of DPP's China Affairs Department and was special assistant to the Vice Minister of Defense Dr. Peter Pi-chao Chen.

21. At least one interlocutor objected to the "newness" of this idea, claiming that Taiwan has always seemed to have had a tactically offensive doctrine.

22. Democratic Progressive Party Policy Committee, "White Paper on Defense," November 23, 1999.

23. Cheng-yi Lin, "The Security of Taiwan in the Year 2000: A Taiwanese Perspective," unpublished paper.

24. Sofia Wu, "Military Chief Explains 'Offshore Defense' Concept," *Taipei Central News Agency*, July 7, 2000.

25. Sofia Wu, "Military Chief Explains 'Offshore Defense' Concept," *Taipei Central News Agency*, July 7, 2000.

26. Sofia Wu, "Military Chief Explains 'Offshore Defense' Concept," *Taipei Central News Agency*, July 7, 2000. See also *Qingnian ribao* [Youth Daily News] (July 8, 2000), p. 3.

27. Huang Ching-lung, Kuo Chung-lun, Hsia Chen, Lu Chao-lung, and Wu Chung-tao, Defense Minister Wu Shih-wen Says: "The Nationalist Forces Now All Know that President Chen Will Not Stand for Taiwan Independence," *Zhongguo shibao*, July 2, 2000.

28. Huang Ching-lung, Kuo Chung-lun, Hsia Chen, Lu Chao-lung, and Wu Chung-tao, Defense Minister Wu Shih-wen Says: "The Nationalist Forces Now All Know that President Chen Will Not Stand for Taiwan Independence," *Zhongguo shibao*, July 2, 2000.

29. "Air Force to Introduce New Strategies," *China Post*, December 27, 2000.

30. Lu Te-yun, "Taiwan to Develop 'Offensive Weapons'," *Lianhebao* (April 16, 1999), p. 1.

31. Chen Shi, "Taiwan: Is It Useful to Develop Offensive Weapons?" *Zhongguo qingnian bao*, December 24, 1999.

32. Alexander Huang, "Homeland Defense with Taiwanese Characteristics: On President Chen Shui-bian's New Defense Concept," presented at the 11th Annual PLA Conference, U.S. Army War College, Carlisle Barracks, PA, December 1–3, 2000.

33. Nien Chen, "An Analysis of the 1996 Republic of China Defense White Paper," *Ch'uan-Ch'iu Fang-Wei Tsa-Chih* [Defense International], July 1, 1996.

34. Taipei Democratic Progress Party (DPP) Policy Committee, "White Paper on National Defense," November 23, 1999.

35. Taipei Democratic Progress Party (DPP) Policy Committee, "White Paper on National Defense," November 23, 1999.

36. 2000 White Paper.

37. 2000 Defense White Paper.

38. Sofia Wu, "Chen Shui-bian Addresses Military Cadets; Urges Loyalty, Defense To Counter PRC," *Taipei Central News Agency*, June 16, 2000.

39. Huang Ching-lung, Kuo Chung-lun, Hsia Chen, Lu Chao-lung, and Wu Chung-tao with Defense Minister Wu Shih-wen: "Defense Minister Wu Shih-wen Says: The Nationalist Forces Now All Know that President Chen Will Not Stand for Taiwan Independence," *Zhongguo shibao*, July 2, 2000.

40. Fang Wen-hung, "Taiwan To Complete Military Restructuring in 3 Years," *Taipei Central News Agency*, February 7, 2001.

41. See Swaine 1999, for details.

42. The *Guofangfa* superseded the 1970 Organic Law of the MND.

43. Cheng-yi Lin, "The Security of Taiwan in the Year 2000: A Taiwanese Perspective," unpublished paper.

44. Peter Yen, "Diversification And Defense Trade Opportunities," U.S. Department of State, September 1999.

45. Office of the Secretary of Defense, *The Security Situation in the Taiwan Strait*, Report to Congress Pursuant to the FY1999 Appropriations Bill, February 26, 1999.

46. "On EW and C3I's Role in Taiwan Defense," *Fang-wei Ta Tai-wan*, November 1, 1995, pp. 354–361.

47. "Tang Fei: PRC May Have Electronic War Supremacy by 2010," *Taiwan Central News Agency*, May 5, 1999.
48. Lu Teh-lin, "Faced with the PRC Information Warfare and 'Electronic Warfare' Threat, Lin Chin-ching Says that Taiwan's Electronic Warfare Capability Remains Inadequate," *Lianhebao*, September 15, 1999.
49. "Taiwan Armed Forces to Test Cable Network in War Games," *Central News Agency*, January 3, 2000.
50. Wen-Hung Fang, "Taiwan Military Using Fiber-Optic Communication System," *Taipei Central News Agency*, May 26, 2000.
51. ABC Radio Australia News, September 30, 2003, can be found at http://www.abc.net.au/ra/newstories/RANewsStories_955464.htm.
52. Office of the Secretary of Defense, *The Security Situation in the Taiwan Strait*, Report to Congress Pursuant to the FY1999 Appropriations Bill, February 26, 1999.
53. 2000 White Paper.
54. "General Confirms Policy on 'Long-Range' Radar," *AFP*, March 23, 1999.
55. The money for the long-range radars was approved in the LY, after an intense dispute over the money for the land. The early-warning radar alone is likely to require at least six years to IOC.
56. Chen Shui-bian, *Xinshiji xinchulu: Chen Shui-bian guojia lantu—diyice: guojia anquan* [New Century, New Future: Chen Shui-bian's Blueprint for the Nation—Volume 1: National Security] (Taipei: Chen Shui-bian Presidential Headquarters, 1999) pp. 50–51.
57. Lin Chin-ching, "Comparison of PRC, ROC Information Warfare Capabilities," March 1, 2000, pp. 68–73.
58. Brian Hsu, "Taiwan Military to Establish Information Warfare Unit," *Taipei Times*, November 23, 2000.
59. Sofia Wu, "Taiwan President Chen Urges Military To Further Upgrade Capabilities," *Taipei Central News Agency*, December 30, 2000.
60. Lin Chin-ching, "Comparison of PRC, ROC Information Warfare Capabilities," March 1, 2000, pp. 68–73.
61. "Taiwan Cites Internet Rumors," *Associated Press*, August 7, 1999.
62. Sumner Lemon, "Hackers Keep Up Attack on Taiwan Web Sites," *Computerworld Hong Kong*, August 24, 1999.
63. "Pro-China Hacker Attacks Taiwan Government Web Site," *Reuters*, August 9, 1999.
64. "Taiwan Cyber-Hackers Strike Back at China," *Reuters*, August 10, 1999.
65. "Taiwan-China Hackers' War Erupts," *Muzi Lateline News*, August 10, 1999.
66. Amanda Chang, "Beijing's 'Information War'," *Taiwan Central News Agency*, August 12, 1999.
67. David Watts, "Virtual Warriors Fire Opening Shots in Cyber Battle," *The Times*, August 18, 1999.
68. Michael Laris, "Chinese Web Warriors: Hackers in Taiwan, China, Trade Shots in Internet Skirmish," *Washington Post*, September 11, 1999.
69. "Taiwan Prepares for Possible Chinese Cyber Attacks," *AFP*, November 2, 1999.
70. Office of the Secretary of Defense, *The Security Situation in the Taiwan Strait*, Report to Congress Pursuant to the FY1999 Appropriations Bill, February 26, 1999.
71. Office of the Secretary of Defense, *The Security Situation in the Taiwan Strait*, Report to Congress Pursuant to the FY1999 Appropriations Bill, February 26, 1999.

72. Lin Chin-ching, "Comparison of PRC, ROC Information Warfare Capabilities," March 1, 2000, pp. 68–73.
73. "Computer Virus Author Revealed," *Associated Press*, April 30, 1999.
74. "Security Ministry of Checking Computer CIH Virus," *Xinhua*, September 24, 1998.
75. Tim Neely and David Cowhig, "China: Information Security," U.S. Embassy (Beijing) report, June 1999. See also "Taiwan College Identifies Virus," *Associated Press*, April 29, 1999; "CIH Virus Culprit Pegged?" *Wired News*, April 29, 1999; and Jeffrey Parker, "Taiwan Virus Suspect Free on Lack of Victims," *Reuters*, April 30, 1999.
76. Office of the Secretary of Defense, *The Security Situation in the Taiwan Strait*, Report to Congress Pursuant to the FY1999 Appropriations Bill, February 26, 1999.
77. Lin Chin-ching, "Comparison of PRC, ROC Information Warfare Capabilities," March 1, 2000, pp. 68–73.
78. Taipei DPP Policy Committee, "White Paper on National Defense," November 23, 1999.
79. Chen Sui-bian, *New Century, New Future: Chen Shui-bian's Blueprint for the Nation—Volume 1: National Security* [xinshiji xinchulu: Chen Shui-bian guojia lantu—diyice: guojia anquan] (Taipei: Chen Shui-bian Presidential Campaign Headquarters, 1999) pp. 74–75.
80. "Taiwan's Information Warfare Task Force," *Jane's Defense Weekly*, May 19, 1999.
81. "Taiwan Steps Up Training to Thwart PRC E-Warfare," *AFP*, September 14, 1999.
82. 2000 White Paper.
83. "Defense Minister Calls for Budget Increase," *China Times*, November 2, 1999.
84. "PRC, Taiwan Modern Warfare Viewed," *Zhongyang ribao*, November 22, 1999.
85. Lin Chin-ching, "Comparison of PRC, ROC Information Warfare Capabilities," March 1, 2000, pp. 68–73.
86. "Military to Test Computer Bugs," *AFP*, August 8, 2000.
87. Fang Wen-hung, "Taiwan Information Warfare Command Established," *Taipei Central News Agency*, January 02, 2001.
88. Brian Hsu, "Taiwan's First Information Warfare Group Enters Service," *Taipei Times*, January 3, 2001.
89. Fang Wen-hung, "Taiwan Information Warfare Command Established," *Taipei Central News Agency*, January 2, 2001.
90. Peter Harmsen, "Taiwan Has 1000 Computer Viruses to Fight Cyber War with China," *AFP*, January 9, 2000.
91. Michael S. Chase, "U.S.–Taiwan Security Cooperation," in James C. Mulvenon, ed., *U.S. Security Cooperation in Asia* (Santa Monica, CA: RAND, 2003).

8

TAIWAN: FROM INTEGRATED MISSILE DEFENSE TO RMA?

Arthur S. Ding

THE CHINESE MISSILE THREAT

In July 1995 and again in March 1996, China launched large-scale military exercises in the Taiwan Strait. In what became known as the Taiwan Strait crisis, China fired several short-range ballistic missiles (SRBMs) at Taiwan and mobilized conventional forces for amphibious attack exercises. The missile launch demonstrated several defense problems for Taiwan. First, China is determined to stop Taiwan's pursuit of international status and to oppose its development of diplomatic relations with other major powers.

Second, Chinese military forces are able to threaten Taiwan's major military assets without physically engaging Taiwanese forces.[1] Taiwanese analysts now realize the real threat to its security is from a missile attack, not a naval blockade, quarantine, and amphibious invasion. The most likely scenario in the view of their defense planners is a Chinese missile attack followed by conventional air and naval attacks.[2]

Third, the missile test exercise demonstrated that Taiwan lacks a viable defense against missile attack. The Chinese are improving the accuracy of their missiles so Taiwan is likely to lose significant forces in an engagement, while the Chinese military suffers very low causalities. Taiwan might not be able to launch an effective counterattack because command and control facilities will have been wiped out. Not only is it likely that Taiwan's major military assets will be destroyed in a missile attack; it will dissuade the United States from intervening in the Taiwan Strait.[3] Taiwan's national security is predicated on the assumption that the longer Taiwan can hold out in the face of a Chinese military invasion, the more likely the United States is to intervene.

One final result of the missile exercises was a psychological one. A missile attack will strike a serious blow to the morale of the general public, which may create popular pressure to accept Beijing's terms for reunification. Uncertainty brought on by political stalemate coupled with the insecurity created by China's missiles is likely to erode confidence in Taiwan's government over the long term.

Taiwan's apprehension has been reinforced by China's ambitious and comprehensive military modernization program designed to achieve information dominance.[4] Apprehension over China's strategic modernization is evident in former Defense Minister Tang Fei's opening remarks at a conference on RMA and defense. In his words, "China will launch six high-resolution satellites next year (2000), and with those satellites, they can watch all of our military actions. They can sit at home and know what we are doing in our backyard."[5] Tang Fei testified in Parliament, "the Chinese military is also doing all its effort to develop 'asymmetrical warfare' by actively developing precision guided ballistic missiles, cruise missiles, information warfare (IW), and electronic warfare (EW) capabilities.... It is estimated that by the year 2005, with the deployment of recently procured weapon systems, China's naval and air capability will exceed that of Taiwan."[6]

Taiwanese defense planners realize that the introduction of missiles by China has changed the nature of a possible war over the Taiwan Strait. Former Defense Minister Wu Shi-wen indicated that China is building a capability for "fighting a local war under high tech conditions" and emphasizing a combined warfare of "high intensity" and "fast pace." Meanwhile, China maintains an operational concept that the "initial battle is the decisive battle." Defense operations in the future will probably be "of short warning time and insufficient defense depth." Minister Wu added that there would be no fixed battle lines and no division between the front and rear. This means war across the Taiwan Strait will be both an issue of armed force and popular morale of the people.[7]

THE U.S. FACTOR

The United States has always been the most important factor in Taiwan's defense and security. The same can be said of Taiwan's development of the Revolution in Military Affairs (RMA): the United States has pushed Taiwan to rapidly undertake RMA-related programs.

The U.S. Defense Department's June 2000 report on China's military power revealed its concern over China's military buildup in the field of IW-related programs.[8] In late June 2003, U.S. government officials cautioned Taiwanese parliamentary members visiting Washington that 2005–08 would be a critical period. The military balance in the Taiwan Strait would tilt toward China and China was likely to attempt to cripple Taiwan's defense in the first wave of strike.[9] The United States requested that Taiwan increase its defense procurement and emphasize Command, Control, Communications, Computer, Intelligence, Surveillance, and Reconnaissance Systems (C4ISR), missile defense systems, long-range radar, and anti-submarine systems.[10] If Taiwan can withstand a Chinese invasion for a certain period of time,[11] China will be deterred and the United States will have sufficient time to intervene. If China perceives that an invasion will fail, peace in the Taiwan Strait can be preserved.

INTEGRATED AIR DEFENSE

Under the shadow of China's missiles and other threats, Taiwan's defense planners set up a task force under the Deputy Chief of General Staff (DCGS) for Operations, the J-3. Headed by former Assistant DCGS for Operations MG Fu Wei-ku, the task force, named "Integrated Air Defense Enhancement Group," was formed in August 1996 and commissioned to study ways to neutralize China's missile threat.[12] Its recommendations, the "Long Term Plan of Integrated Air Defense," were approved by Chief of the General Staff, Gen Tang Yao-ming in April/May 1999.[13] They have been expanded to represent Taiwan's version of the RMA.

The "long-term plan" comprises several elements.[14] With regard to the missile threat, the plan proposes active (*ji ji fang yu*) and passive (*xiao ji fang yu*) defenses. The active component involves use of a missile defense system to intercept missiles in the atmosphere; the passive component involves hardening major military assets and shelters to limit the damage caused by a missile attack.

The plan considers both ground- and sea-based lower-tier missile defense systems. The task force recommended procurement of the Aegis combat system and the Patriot system along with the indigenously developed Sky Bow (*tien gong*) anti-tactical ballistic missile and air defense system to protect major political and military assets. These recommendations do not rule out the option of developing "defensive counter systems"(*fang wei xing fan zhi wu qi*), with sufficient range to cross the Taiwan Strait.

Economic considerations are a major factor behind the option of developing the "defensive counter system." Taiwan's defense planners estimate the cost ratio between offensive and defensive missiles to be roughly between 1:9 and 1:18. It is too costly to develop a missile defense system. Yet due to rapid economic development, China's southeast region has become an inviting target. If Taiwan developed a "defensive counter system," China would be forced to divert resources into defensive systems.[15]

The second element in the long-term plan involves early-warning systems and reconnaissance capability. The task force proposes to procure long-range radar to increase time for warning and response. The long-range radar can also cue missile trajectory information to the combat radar. The task force also recommended that Taiwan procure satellites to conduct reconnaissance and surveillance.[16]

The third element involves buildup of an integrated command and control system among the three services to combine all air defense units into one efficient system. This requires developing a Taiwanese (or low-end) version of C4ISR to buttress the integrated air defense system.

The early warning systems, reconnaissance capability, integrated command, and control system, along with other measures, form the core of Taiwan's version of information warfare (IW). Taiwan's defense planners realize that IW occupies the highest strategic position for overall security in the Taiwan Strait.

Evidence suggests that this plan is being implemented. Taiwan made a procurement request for Aegis-equipped destroyers to the United States in the fall of 2000, although the newly elected Bush Administration postponed the procurement request in April 2001. Taiwan also requested the PAC-2 plus system, which has been delivered. This system is to be supplemented with the indigenously developed Sky Bow I and Sky Bow II missiles.

Taiwan is reported to be aggressively developing a "defensive counter system." Two types of missile programs are underway. The first, code-named Di jing, aims at developing a ballistic missile with mid-range capability, between 1,000 and 2,000 kilometers.[17] This program is expected to be completed before 2005, when many U.S. defense experts predict Taiwan may lose military superiority over China.[18] The ballistic missile program faces several challenges, the major one being increasing accuracy, reducing deployment time, and maximizing response capability without using a mass destruction warhead. Some subsystems have reportedly been tested, but it is too early to conduct the whole range of flight tests.

The second missile program, code-named "Hsiung Feng 2E," involves development of a subsonic land-attack cruise missile (LACM) with a range of 1,000 kilometers, capable of hitting military targets in China's Nanjing, Jinan, and Guangzhou Military Regions. Media reports indicate the design of the cruise missile is based on the U.S.-made Tomahawk cruise missile.[19]

Taiwan is pursuing two early-warning systems. The first is a reconnaissance satellite. As a result of cooperation with Israel, Taiwan can use the Israeli EROS (Earth Resource Observation Satellite) to obtain real-time black/white photographs with a resolution of 2 meters. The EROS-1 is reportedly able to detect Chinese military systems including the M-9/M-11 ballistic missile. The present contract with Israel allows Taiwan to have full control of photo-taking as the EROS satellite enters a 1,000-kilometer radius of Taiwan's satellite receiving station.[20]

Currently, Taiwan obtains information from a pool of reconnaissance satellites. In addition to EROS-1 and the following EROS series of satellites, Taiwan buys images from a U.S. commercial satellite, the IKONOS (with a resolution of 1 meter), the French Spot satellite, and the Taiwan-assembled Hua Wei-2 (with a resolution of 2 meters). By 2005, Taiwan should be able to obtain sufficient information from those satellites.

For long-range radar, the United States agreed to sell the powerful AN/FPS-115 PAVE PAWS phased array radar in the spring of 1999. Taiwan's warning of missile attack will increase by 6 minutes and Taiwan will be able to detect a target several thousand miles away. All of Taiwan's air defense systems will be integrated with this radar.[21]

Efforts are underway to establish an integrated command and control system. The program, code-named "Bo Sheng" (Fight to Win), aims at establishing a data link to integrate the command and control systems of the three services, including their air defense systems. Currently, the navy and air force have their own air defense systems, code-named "Da Cheng" (Great Success) and "Qiang Wang" (Strong Net) respectively, which are operated independently of one another. Integration will also extend to the newly developed unmanned aerial vehicle (UAV) and underwater surveillance systems.[22]

A new communication system for the army has been set up as well. Media reports indicate that the army procured the first Improved Mobile Subscriber Equipment (IMSE) system from the United States in August 1998 to build up army-wide tactical communications. It is expected that the army will procure two more units to establish an island-wide army communication system.[23]

Under the Bo Sheng program, the army has organized a battalion-level electronic warfare (EW) system. This unit has reportedly participated in military exercises, launching EW attacks against the "red team's" air defense radar. The army will upgrade its four "communication units" to "communication and information units" with the capability to conduct a full range of EW missions. The navy and air force also have their own EW units.[24]

Developing an IW Capability

Since 1998, developing an IW capability has become a priority for Taiwan's military. In addition to acquiring C4ISR and early-warning capabilities, Taiwan's defense planners have undertaken other measures toward this end.[25] First, organizationally, a "Guiding Committee for Information Warfare Strategy Development" was formed atop a group of related units involved in IW development to direct the overall IW effort. Below this guiding committee, a special agency was designated to handle the overall development of IW for Taiwan's military. The Communication Electronic and Information Bureau (CEIB) has full responsibility for this job and wideranging influence over the buildup of Taiwan's IW capability.

Second, the Taiwanese have set up an "IW lab" charged with the responsibility for research into and planning of advanced IW technology. The Chungshan Institute for Science and Technology reportedly plays the role of the lab.[26] A "Central Emergency Response Team" has been formed to handle various emergent matters through cyber monitoring and rapid response.

Third, a newly formed "Communication and Information Command" or so-called "IW troop" directly subordinate to the Chief of General Staff, has the mission of undertaking research on offensive and defensive IW tactics, monitoring the Internet and Intranet, and launching "defensive counter actions" during wartime to facilitate air superiority, sea denial, and antilanding warfare.[27] This unit, established on January 1, 2001, also called the "Tiger Team," has reportedly participated in an annual large-scale military exercise.[28]

Fourth, the Taiwanese formed a coordination mechanism called the "joint working group," composed of members from the Ministry of National Defense (MND), National Security Council, Ministry of Finance, and intelligence units. The group is commissioned to study the impact of IW patterns. There is also an unnamed unit responsible for collecting, analyzing, and disseminating information and intelligence as well as coordinating operations by services and arms.[29]

Fifth, Taiwan established a development plan for IW. Former Defense Minister Tang Fei pointed out to Parliament that in the near term, Taiwan was emphasizing the strengthening of information protection mechanisms. The mid-term goal is to build EW software and hardware capabilities through system integration, improve the technology of frequency management, and strengthen electronic detection. The long-term goal is to improve research and development (R&D) of technology and tactics continually so that a digitized force can be formed and an IW capability with both offensive and defensive functions can be established.[30]

The range of initiatives enumerated above demonstrates that the development of an IW capability has become a priority for Taiwan's defense planners. This trend is reflected in the 2001 defense budget. "A defense official said that the focus of the military construction in the next one and two years will be placed on the integration of C4ISR, data link, EW, IW, early warning, protection measures against the electromagnetic pulses (EMP) bomb, and defensive counter systems. Of the NT$47.3 billion of capital investment, those systems will take the lion's share."[31]

Former Defense Minister Wu Shi-wen echoed this focus in his testimony to Parliament. Priorities for capital investment for fiscal year 2002 (from January 1 through December 31, 2002) would be those which are already being procured: the C4ISR system, "defensive counter systems," and those subsystems/components that are in severe shortage.[32]

There are several reasons why Taiwan has given precedence to the development of an IW capability. First, China's rapid development of IW capabilities is perceived as an imminent threat.[33] As former Defense Minister Chiang Chung-ling discussed in his statement to Parliament, Chinese military forces from the General Staff Department down to the Military Region, Group Army, and major surface combatants all have deployed electronic reconnaissance and jamming capabilities. China's air force already has more than ten EW aircraft able to conduct jamming and intelligence gathering functions.[34]

The second reason for Taiwan's efforts to develop an IW capability is its comparative advantage in information technology (IT). Tang Fei's report to the Legislative Yuan reflected this confidence: "We already have an institutionalized information education system; good basis has been laid on information construction; our private sector already has done well in the field of anti-virus software and internet protection products."[35]

Perceived low cost is the third reason for emphasizing IW capabilities. IW is seen as a low-cost tool to help Taiwan balance defense needs with economic development, and acquire asymmetric means of waging war against China. Former Defense Minister Tang Fei, in a speech to the general public, emphasized that to accomplish its objective of countering a larger opponent and, at the same time, saving resources, it is necessary to make the development of an IW capability the top priority.[36]

ORGANIZATIONAL CHALLENGES AND RESPONSES

While pursuing the RMA is highly desirable, a variety of factors will inhibit Taiwan's efforts. The first involves lack of integration between Taiwan's

RMA program and its overall military strategy. Coordination between end users, the three services, and suppliers has been poor.[37] The MND's IW Guiding Committee is loosely organized. Its officers do not have expertise in the IW field and are not involved in IW matters full time.

Below the top brass, each service lacks a specialized agency to study service needs. The CEIB alone has the expertise needed to plan Taiwan's approach to IW, but it is an agency completely composed of technical staff and it lacks experience in comprehensive defense planning and military operations. The CEIB's proposals do not address the larger question of how IW should be integrated with overall operational planning and military strategy. Lack of effective coordination for developing an IW capability reflects a larger problem: an inordinate emphasis on the development of hardware without considering how organization and doctrine should make concomitant changes.

Organizational restructurings during 2002 and after might help address the lack of coordination between technology, organization, and doctrine. Taiwan's military system began two series of organizational reforms in the mid-1990s. The first round streamlined the high command and consolidated force structure.[38] Formally launched in mid-1997 through June 2001, it was unrelated to the RMA; rather, it was a reaction to unfavorable domestic circumstances. The second round focused on transforming civil–military relations and was launched when calls for civilian control arose and the RMA had became a popularly accepted idea.

Several factors led to the streamlining of the high command. First, the increasing cost of social welfare programs due to democratization and the need for infrastructure investment strained financial resources and forced a reduction in the defense budget. Second, huge military personnel costs, together with operational costs and capital investment contributed to cuts in force structure.

Third, the decline in available draftees reduced force structure. Fewer skilled young people were interested in serving in the military when they could be better off financially working in the private sector. As Taiwan has proceeded toward democratization, the military's authoritarian culture has also become less attractive to the younger generation. The trend has reversed itself in the last two years, but only due to Taiwan's anemic economy. Declining population growth is also a problem. Lower birth rates, the result of increasing living pressure and the declining economy, as well as the outflow of population due to a rising crime rate and better educational opportunities abroad, have reduced the pool of available draftees.

Finally, China's rapid military modernization program contributed to streamlining. Since the Gulf War, China has rapidly modernized its military force in order to fight a "local war under high-tech conditions."

Taiwan's military reorganization of the mid-1990s, called the Jing Shi program, consisted of two parts: streamlining at the command level and consolidating the field levels. The hope was that upon completion of the Jing Shi program, total manpower would be reduced while firepower, mobility, and mobilization capability would be increased. Twenty-five departments in the General Staff under the five DCGS were reduced to sixteen. At the General Political Warfare Department, five departments were reduced to four. Several arms training schools merged into one for each service, while several military colleges were merged to form the National Defense University. Changes were implemented in each of the services and at the Tri-Service Joint Logistic General Headquarters. The Jing Shi Program also established an Education, Training, and Doctrine Development Command in each of the services and merged two agencies to establish the CEIB.

With regard to consolidation, the biggest structural change occurred in the army where division-level units were gradually replaced with combined arms brigades. Five types of combined arms brigades were organized: armored infantry, airborne cavalry, motorized infantry, special operations, and tank. The five combined arms brigades will become a striking force with strong mobility and firepower.

Although significant changes have been occurring, many questions were not addressed in the streamlining and consolidation efforts. There was no real sense of organizational reengineering as proposed by RMA advocates. The end result was the establishment of a percentage goal for across-the-board cuts without consideration for changes in military strategy. No discussion or debate took place over how the consolidation of the army force structure should come about[39] or how many combined arms brigades are needed. No experimental formation of brigades was considered before the formal decision was made.[40] The concept of joint operations among the three services was also neglected.

POLITICAL TRANSFORMATION AND CIVIL–MILITARY CHANGES

More far-reaching changes began in the early 1990s in response to calls for greater civilian control over military affairs. Taiwan lifted four decades of military rule and abolished regulations governing mobilization for recovering mainland China.

During its first 40 years, real military command power rested in the hands of the Chief of the General Staff (CGS), who reported directly to the President. The Minister of National Defense played a nominal role under the Premier. The CGS enjoyed tremendous power without accountability, while the powerless minister had to answer to parliament.[41]

The National Defense Law[42] redressed this imbalance of power. It stipulates that the defense minister must be a civilian and places the CGS and the entire staff under the institutional authority of the minister. The CGS now serves as both the military staff for the minister and commander of military operations under the minister's supervision. The CGS is also subject to oversight by parliament.

To consolidate the minister's authority over defense policy, the functions and size of his staff have substantially increased. New departments have been established while others have been moved out of the General Staff Head-quarters (GSH). The Strategic Planning Department, Integrated Assessment Office, Resource Department, Armament Bureau, and Mobilization Department were established, while the Personnel Department, Bureau of Budget and Accounting, and Military Justice Bureau were transferred out of the GSH.

The defense minister's control over the three services has also been enhanced. The new defense law places the three services under his command. The services focus only on training and doctrinal development. Operational power during wartime is relegated to the CGS. The new defense law also created a new division of labor within the military system. Previously, the GSH possessed all powers ranging from defense policy planning to military operations. The law divides defense into three subfields—defense policy, armament, and military operations—to facilitate specialization and enhance overall military capability.[43]

The implications of these changes in civil–military relations and of consolidation of the defense minister's authority for the RMA are significant. Previously, there was almost no interaction between uniformed personnel and the civilian sector, and new ideas could not find an easy foothold in the system. The lifting of martial rule and increased democratization has increased military professionalism and left domestic order to the civilian justice sector.

The new structure of civil–military relations, based on a participatory system instead of an authoritarian regime, should also bring new inputs into the military that will facilitate change.[44] Mechanisms for integrating the military and civilian sectors have been devised. Under a pilot program, effective from April 2003, the MND and Ministry of Foreign Affairs have exchanged staff. This program is expected to expand to other areas after a six-month experimentation period.

Integration at higher levels is also occurring. There have been regular meetings between the Ministers of National Defense and Foreign Affairs. The National Security Council of the Presidential Office is playing a more active coordination role among the ministers that deal with national security policy.[45]

Finally, the new law requires the office of the Minister of National Defense to recruit civilians, who must comprise at least one-third of the staff in the office of the defense minister. Uniformed staff must be rotated every 2–3 years, which has been detrimental to the accumulation of professional expertise.

IMPROVING THE QUALITY OF PERSONNEL AND CIVILIAN–MILITARY INTERACTION

Taiwan's military is experiencing a major improvement in the quality of its personnel, the product of changes in the economy, which provide a solid foundation for military efforts to develop Taiwan's version of RMA.

Previously, Taiwan's military was composed of two types of personnel: professional career officials and two-year conscripts. Career officials, all graduates of military academies, have dominated the military since 1949.[46] With improvements in the economy during the country's first four decades, young talented people opted for promising opportunities in the private sector.

The consequences for Taiwan's military were serious. The quality of leadership deteriorated; the military became an isolated enclave; a rigid and authoritarian culture flourished; innovation was not encouraged; and promotions were tied to seniority and personal relationships. The result was absence of internal momentum for reform and modernization.

A weakening economy reversed this trend. Young, talented people have begun to pursue military careers, and the military academies have recruited sufficiently high-quality students.[47] A Reserve Officer Training Corps (ROTC) system established in 1997, modeled on the American system, has attracted many talented civilian students.

The conscript system has also gradually shifted to a volunteer system.[48] There is a growing consensus that a voluntary system has advantages over a conscription system given the demands of operating expensive high-tech weapons and the length of time needed for soldiers to achieve competency in operating these systems. A voluntary system facilitates the accumulation of necessary experiences, while better salaries attract high-quality individuals.[49]

Defense Minister Tang Yao-ming has pointed out that in the future, the standing combat troops, and combat support and logistics units will be composed primarily of volunteer soldiers, while the reserve force will be mainly composed of conscripted soldiers. By 2011, Taiwan's military will be predominantly a volunteer force.

Defense planners have come to recognize the importance of interaction between the military and civilian sectors. Policy now encourages military personnel to pursue advanced studies at civilian colleges and universities, and financial assistance for this is a regular defense budget item. Allowing military personnel to pursue advanced studies at civilian institutes will help the defense ministry, be more cost effective, and allow soldiers to prepare for careers in other sectors of society. This policy has increased the interaction between the civilian and military sectors. Many uniformed personnel have been admitted to programs in civilian institutes, including strategic/security studies, public policy/administration, and MBA/EMBA programs. Interaction in educational institutes allows uniformed people to keep abreast of developments in modern society and to learn about civilians sectors and approaches.

ECONOMIC ENABLERS AND INHIBITORS OF THE RMA

Taiwan has been a leading supplier of IT hardware products in the world, including memory chips, keyboards, mother boards, PCs and portable PCs, monitors, mouses, and mobile phones. The U.S. Department of Defense has pointed out that Taiwan is known for its computer/network security anti-virus and firewall programs. Taiwan's IT industry is also developing computer games. All these serve to deflate pressures on Taiwanese investors to move IT production lines to China and increase incentives to take advantage of Taiwan's liberalized society. Although these developments lay the basis for Taiwan's military to move into the field of IW, Taiwan's IT industry has not contributed much to the country's defense. There are several explanations for this.

First, Taiwan produces mostly original equipment manufacturer (OEM)-oriented products in accordance with foreign buyers' purchasing orders.[50] Most of these are low-end products without original R&D input. Second, Taiwan has no economic incentive to develop IT products to military specification. The domestic market for military technology is small, and diplomatic isolation hampers exports. Third, a risk-avoidant mentality toward military procurement hinders indigenous development of IT products. Taiwan's military procures systems from advanced countries because defense planners and the services lack confidence in indigenously developed systems. Fourth, there is no government agency responsible for development of the defense industry,[51] so there is no regular connection between the defense and commercial sectors. Fifth, the time it takes to field a sophisticated program contradicts the bureaucrats' emphasis on short-term performance to ensure promotion. Many high-ranking military leaders prefer a "quick fix," rather than long-term solutions, and this is detrimental to the development of Taiwan's RMA.

Though Taiwan has developed a well-known IT industry, its contributions to the RMA and IW have been limited to cyber virus/anti-virus and hacking programs. Taiwan continues to rely on the United States to provide C4ISR-related systems.

AN RMA WITH TAIWANESE CHARACTERISTICS?

Exploiting the RMA requires a forward-looking strategy. Based on the potential military application of new ITs in the next 10–20 years and a new mode of warfare/operations, a military system can be transformed in its organization and operational doctrine through a continuous process that poses continual challenges.

If one accepts the above RMA definition, then what Taiwan has accomplished over the past several years cannot be interpreted as an RMA. Taiwan has reacted to unfavorable circumstances, including declining defense budgets, a weakening economy, decreasing numbers of draftees, and a rising Chinese military threat. This does not represent a proactive or visionary strategy.

The military reluctantly accepted the RMA concept only when the civilian sector entered the discussion.[52] Some officials discounted the concept entirely. Though an ad hoc group was formed at the Armed Force University (later, National Defense University) to study the RMA, their study has not received unanimous support from defense planners and has had minimal impact.[53] What is particularly distinctive is the role played by the United States. Concerned with stability in the Taiwan Strait and its credibility, the United States has persistently and actively engaged in Taiwan's RMA program[54] and pushed Taiwan hard to undertake RMA-related programs.[55]

Yet Taiwan's military has not been able to indigenously develop and leverage IT. It lacks internal mechanisms to support innovation and expertise in strategic planning.

Under the circumstances, Taiwan's RMA will be IW-centered, heavily reliant on computer hacking and virus/anti-virus-related offensive/defensive capability; a linked command, control, and communication system; and limited reconnaissance and counter-strike capability. There will be no sensor-to-shooter and precision-attack capability, or digital force/unit like the one that launched cruise missiles at Saddam Hussein 15 minutes after intelligence was gathered.

Taiwan knows that it needs an early warning and reconnaissance capability for strategic warning. Procuring long-range early-warning radars and receiving satellite imagery information from Western satellites are necessary for enhancing its strategic warning capability. However, it is unlikely Taiwan will develop real-time on-line capability.

Because it lacks capable hard-kill systems such as long-range precision-strike capability, the EMP bomb, and the anti-radiation missile, Taiwan will likely rely on an Internet-centered IW capability such as computer hacking and logic-bombs to provide it with the offensive capability to disrupt and destroy China's Internet-based command and control system before a war breaks out.

Taiwan is likely to build up a C4ISR capability through a data link system with the assistance of the United States. The United States plays the most important role in pushing Taiwan to pursue an RMA-related program. Since the 1996 Taiwan Strait crisis, the United States has increased security engagements with Taiwan.[56] Through the data link system, Taiwan will be able to integrate the command center with the command posts of the three services, as well as the major platforms such as jet fighters, early warning jets, warships, and various surface-to-air missile stations.

It is also likely that Taiwan will build up a limited "defensive counter system" capability to demonstrate its resolve against a Chinese invasion. Fully aware of the sensitivity of this issue, Taiwan may equip its defensive counter system with a highly explosive or fuel air explosive warhead[57] instead of a weapons of mass destruction (WMD) warhead.

CONCLUSION

The RMA has become a fashionable topic in Taiwan. Taiwan's military has tried to emulate advanced countries, particularly the United States, in order to modernize its military and deter or repel a Chinese invasion. However, these changes are more easily discussed than implemented.

Taiwan lacks many of the conditions for an RMA as discussed in the introductory chapter to this volume. In the political field, Taiwan's participatory democracy is conducive to encouraging creativity and innovation. A gradual shift to civilian control from the previous "stove-piped" system allows new ideas to be more easily introduced into the military. Civilian control and organizational restructuring makes coordination among strategic planning, assessment, resource allocation, and force structure more rational and allows uniformed personnel to focus on training and doctrinal development.

However, Taiwan possesses a weak state structure. The transition to participatory democracy has created "pork-barrel" politics: politicians appeal to welfare programs to gain votes in elections and the defense budget has declined.

Political transformation allows more debate on the previously prohibited "Taiwan Independence" appeal. However, greater tolerance of debate has produced an identification rift in society and the military. An argument that undermines support for military modernization surfaced: political differences are the source of cross–strait tensions and military modernization is not the best solution.[58]

Taiwan's economic strengths in the IT area have not translated into any advantages in developing the RMA. Taiwan is good at putting mature technology into the production line, but R&D still occurs in Western countries. OEM type production has not assisted Taiwan in developing its RMA.

The country's small economic scale and diplomatic isolation constitute serious barriers for investment by Taiwanese businesses. It requires significant investment to transform a dual-use commercial product into a product that meets military specifications and Taiwan's small defense market, coupled with diplomatic isolation, which curtails exports, makes military investment unviable. Taiwan's weakening economy and declining defense budgets due to diminishing tax revenues and pork-barrel politics have seriously constrained Taiwan's investment in the RMA.

Taiwan's society has undergone changes that bode well for RMA efforts. Five decades of integration with the world and increasingly liberalized politics have produced a climate that tolerates and encourages individualism, and is conducive to creativity and innovation. A highly information-literate population and the popularity of IT products in society have given rise to the development of cyber-warfare capabilities, including computer hacking and virus/anti-virus capability.

Yet there are still social characteristics that stifle innovation and creativity. School exams are memory and computation oriented. Students are forced to practice redundant computational skills for mathematics and the natural sciences. The education curriculum is designed to promote moral grounding and pass exams, not to help students learn how to solve problems.

Changes within the military have been favorable to the development of the RMA. The weakening economy and rising unemployment have made a military career more attractive to the younger generation. Although young recruits may breathe new life into the military, the rigid management culture may frustrate newly recruited young people who entered the military for job security, good pay, and with the expectation that war would be unlikely.

Organizational restructuring in the military should support forward-looking planning, which is necessary for implementing an RMA. New agencies, such as the Strategic Planning Department, Integrated Assessment Office, and Resources Department are expected to make the management of strategic planning, force structure, and military strategy more logical in the long term. The recruitment of civilian staff into the office of the Minister of National Defense should increase professional expertise.

Yet amid these positive factors, inhibiting forces remain. Service rivalry continues to be strong. There have been numerous delays in resource allocation for building a C4ISR capability. A risk-averse nature has hindered development of indigenous expertise. Lack of a persistent policy supporting and encouraging development of indigenous expertise has created "rent-seeking" behavior, benefiting the individual's career at the expense of the nation.

Table 8.1 summarizes the factors enabling and inhibiting transformation within Taiwan's military. There is potential for Taiwan to accomplish a limited RMA, but the United States will be the most important factor. The United States provides much of the necessary technology and expertise. America's concern with stability in the Taiwan Strait, and with its own credibility in the region, explains why the United States is pressing Taiwan hard to modernize its military. In the end, the degree to which Taiwan is likely to succeed will depend upon the United States.

Table 8.1 Drivers, enablers and inhibitors of Taiwan's RMA

Drivers	Chinese military threat; Military Reform; U.S. pressure	
	Enablers	Inhibitors
Political factors	Liberal political system Civilian control	Political identification Weak state structure Diplomatic isolation
Economic factors	Expertise in software development	Lack of indigenous core technology Small domestic market Weakening economy Declining defense budgets
Social and cultural factors	High information literacy Popularity of IT products Open society Integration with global economy Toleration of individualism	Strict hierarchical order Inadequate education
Military factors	Professional army Improved personnel quality Re-division of labor	High inter-service rivalry Risk-averse attitude Overly rigid discipline Rent-seeking mentality Technology-oriented

NOTES

1. *China Times*, January 14, 2001, p. 1. Taiwanese defense planners are concerned that Taiwan cannot withhold Beijing's initial waves of attack on command and control systems, airports, ports, and radars, particularly in the case of a missile attack. Therefore, there are calls for hardening major military assets.
2. *China Times*, January 9, 2000, in <http://forums.chinatimes.com.tw/special/taiwanstrait/89109f01.htm>.
3. *China Times*, January 14, 2001, p. 1.
4. Mark Stokes, *China's Strategic Modernization: Implications for the US* (Carlisle, PA: US Army War College Strategic Studies Institute, September 1999). This book has been translated into Chinese and widely distributed in Taiwan's military.
5. *Xin Xin Wen* [*The Journalist*] (Taipei), no. 628 (March 13, 1999).
6. Tang Fei, *Defense Report by the Ministry of National Defense to the Committee of National Defense*, the Legislative Yuan, March 24, 1999, pp. 5–6.
7. Wu Shi-wen, *Defense Report by the Ministry of National Defense to the Committee of National Defense*, the Legislative Yuan, September 27, 2000, pp. 3–4.
8. Office of the Secretary of Defense, *Annual Report on the Military Power of the People's Republic of China*, Report to the Congress Pursuant to the FY2000 National Defense Authorization Act, June 22, 2000.
9. The Bush administration approved a package of arms sale items in April 2001, including four Kidd-class destroyers. The Kidd program encountered strong opposition in Parliament and was not approved until early 2003. For related debates and MND's responses, see <http://www.ettoday.com/2003/01/10/91-1398709.htm>, and <http://www.libertytimes.com.tw/2002/new/oct/ 17/today-p2.htm>.
10. *CT*, June 26, 2003, p. 4, and July 2, p. 6, as well as *UDN*, June 30, 2003, p. 4.
11. A U.S. official reportedly was to inform Taiwanese parliamentary members that Taiwan has to defend itself for at least two weeks before a U.S. intervention, *UDN*, June 27, 2003, p. 6.
12. The task force was still headed by Fu Wei-ku after he was promoted to lieutenant general, and became the commandant of the Command and Staff College of the Air Force. After the study had been completed, the task force was renamed the "Task Force for Integrated Air Defense and Integrated Counter Measures," and was directed by the DCGS for operations. *United Daily News* (Taipei), June 18, 2001, p. 2.
13. *China Times* (Taipei), September 21, 1999, p. 4.
14. Unless otherwise cited, the following information is drawn from *China Times*, September 21, 1999, p. 4.
15. *United Daily News*, June 19, 2001, p. 4.
16. As early as May 1996, Taiwanese defense planners have thought it necessary to develop satellite early warning and reconnaissance capabilities, along with long-range radar. *China Times*, May 18, 1996, p. 1.
17. *United Daily News*, June 18, 2001, p. 2. There are strong voices in Taiwan advocating development of ballistic missiles as retaliatory weapons. Estimates are that in ten years Taiwan will be able to develop a missile with a range of 2,500 kilometers. *China Times*, January 9, 2000, in <http://forums.chinatimes.com.tw/special/taiwanstrait/89109f01.htm>.
18. U.S. Department of Defense, *The Security Situation in the Taiwan Strait*, in <wysiwyg://28/http://www.chinatimes.com.tw/report/tmd/news/report1.htm>, August 2, 1999.

19. *United Daily News,* June 26, 2001, pp. 1; 3.
20. The satellite information is from *United Daily News,* August 12, 2001, p. 1.
21. "Taiwan will procure new radar system," in <http://www.future-china. org.tw/csipf/press/digest/dgst 891120.html>.
22. Unless otherwise cited, the Bo Sheng information is drawn from *China Times,* June 10, 2001. It is reported that the Bo Sheng contains subprograms to cover different missions, including an EW system for the army, data link for the three services, and communication equipment for the army. This author thanks Der-yeun Lu of *United Daily News* for providing this source.
23. *Central Daily News,* January 23, 2000, p. 4, and *Taipei Times,* December 26, 2000, in <http://www.taipeitimes.com/news/2000/12/26/story/0000067028>.
24. *China Times,* June 10, 2001.
25. Unless otherwise cited, all related information is drawn from Tang Fei, Defense Report to the Committee of National Defense, the Legislative Yuan, November 1, 1999, pp. 8–9.
26. Media report that the Chung Shan Institute has a "Net Security," or the CSII (Chung Shan Information Infrastructure) program that is designing a protection system for the intranet systems of the three services. *The Journalist,* no. 628 (March 18, 1999). This report demonstrates that the Chung Shan Institute plays an important role in IW capability development.
27. Information on the IW troop is drawn from Wu Shi-wen, Defense Report by the Ministry of National Defense to the Committee of National Defense, the Legislative Yuan, March 5, 2000, p. 9.
28. It is reported that beginning in 1998, Taiwan began to organize the IW troop, and priority for procurement shifted from hardware to software. See *Global Times* (Beijing), March 23, 2001, from <http://gptaiwan.org.tw/~cylin/china/2001_3_23.htm>.
29. *Global Times,* March 23, 2001, from <http://gptaiwan.org.tw/~cylin/china/2001_3_23.htm>.
30. Tang Fei, Defense Report by the Ministry of National Defense to the Committee of National Defense, the Legislative Yuan, November 1, 1999, p. 10.
31. There were different reports for this figure but they all pointed to the trend of developing IW capability as the priority. See *United Daily News,* September 3, 2001, in <http://be1.udnnews.com/2001/9/3/newsfocusnews/focus/445428. shtml>, and *China Times,* September 6, 2000, p. 4.
32. Wu Shi-wen, Defense Report by the Ministry of National Defense to the Committee of National Defense, the Legislative Yuan, October 3, 2001, p. 19.
33. Tang Fei, Defense Report by the Ministry of National Defense to the Committee of National Defense, the Legislative Yuan, November 1, 1999, pp. 7; 9.
34. *United Daily News,* May 16, 1998, p. 13.
35. Tang Fei, Defense Report by the Ministry of National Defense to the Committee of National Defense, the Legislative Yuan, November 1, 1999, p. 9.
36. *Liberty Times,* July 25, 1999, p. 2.
37. Personal interview with defense officials, January 9 and 15, 2002.
38. This section is heavily drawn from Arthur S. Ding and Alexander C. Huang, "Taiwan's Military in the 21st Century: Redefinition and Reorganization," in Larry Wortzel, ed., *The Chinese Armed Forces in the 21st Century* (Carlisle, PA: U.S. Army War College Strategic Studies Institute, December 1999) pp. 253–283. For the purpose of this chapter, only that portion of the streamlining and consolidation are cited.

39. Consensus on reducing the size of the strategic unit for the army from the division level has been reached, but no consensus on the replacement was made.
40. Because of a lack of an experimental formation of the combined arms brigade before the decision was made, the unit type has met with problems. The streamlining program abolished the coordination officials from navy and air force at the division level, making joint warfare impossible for the combined arms brigade; combat support units are dispersed, reducing firepower capability.
41. For an excellent analysis of this imbalance power relation, see Michael Swaine, *Taiwan's National Security, Defense Policy, and Weapons Procurement Processes* (Santa Monica, CA: Rand, 1999).
42. The law was ratified by parliament in January 2000, and formally put into effect in March 2002. There were two years of preparation for transition.
43. Some military officials working at these new agencies said they have been very busy because the defense minister instructed them to conduct many studies designed to push new policies. This might imply that the re-division of labor has generated specialization.
44. Ching-pu Chen, Idea and Vision of Taiwan Defense Organization Reform, unpublished paper, May 2002. Ching-pu Chen was one of many who provided substantive advice to the draft of the new defense law.
45. *CT*, July 13, 2003, p. 6.
46. More specifically, the highest rank positions were all dominated by those from military academies of the three services, while those from professional academies retired from the military at the colonel level.
47. In 2003, the admission rate for those applying to military academies was on average 5 percent. *UDN*, July 20, 2003, p. 3.
48. The MND set up a special center to undertake recruitment. For a report on the strategy, goal, tactics, and measures for the recruitment system, see Guo jun ren cai zhao mu ce lue yan jiu [Study on the Recruitment of Human Resource in the Military], in <http://military.fhk.edu.tw/military/research/seminar/combine/國軍人才招募策略之研究.html>.
49. For parliamentary members' opinion, see <http://www.epochtimes.com/b5/2/4/7/n182089.htm>, and for a related opinion poll, see <http://vote.sparklit.com/poll.spark/615084> and <http://www.yok.idv.tw/home2-article.asp?channel=B&serial=219>.
50. OEM, refers to a company which repackages equipment by other companies. OEM may not add anything except their name to the product. In some cases, they integrate components into complete systems.
51. Beginning in the 1990s, there was speculation in the United States that Taiwan was attempting to build closer security relations with the United States through its military procurement policies. After the de-recognition of foreign relations with the United States in 1979 and the August 17 Communique of 1982, Taiwan established a policy of self-reliance through technology transfer. In the 1990s self-reliance was replaced with procuring arms from abroad. See Bates Gill and Richard A. Bitzinger, *Gearing up for High-Tech Warfare? Chinese and Taiwanese Defense Modernization and Implications for Military Confrontation across the Taiwan Strait, 1995–2005* (Washington, D.C.: Center for Strategic and Budgetary Assessment, 1996).
52. A typical instance is that the Division for Strategic and International Studies (DSIS) of the Taiwan Research Institute sponsored a conference on RMA and defense in April 1999.

53. Personal interview with defense official, January 24, 2002. It was likely that the physical threat posed by China's missiles during the 1995/1996 Taiwan Strait crisis shocked defense planners and the general public, while the RMA is an intangible concept, although one with wider ranging repercussions. This influenced defense planners' responses to the recommendations of the Integrated Air Defense Group and the RMA study by the Armed Force University.
54. For the U.S.–Taiwan security engagement, see Arthur Ding, "U.S.–Taiwan and U.S.–China military relations after the 1996 Taiwan Strait crisis," in Yuh-woei Wang, ed., *Bush Administration's China Policy and the Development of the Cross–Strait Relations* after the 2001 APEC Meeting (Taipei: Prospect Foundation, December 2001) pp. 2–30. Former Pacific Commander, Adm. Dennis Blair was sent to Taiwan in April 2003 to observe how Taiwan would withstand China's first strike and to assist Taiwan in assessing the annual joint military exercise, The *Han Guang 19. Chinatimes*, July 10, 2003, p. 2.
55. The best example was the remarks about China's military threat made by U.S. DoD officials to Taiwanese parliamentary members in June 2003. *CT*, June 26, 2003, p. 4.
56. Arthur Ding, note 54.
57. China reportedly has equipped their M-9/M-11 missiles with fuel air explosive warheads. See <http://www.ettoday.com/2002/11/14/706-1375490.htm>.
58. Denny Roy, "Taiwan's Threat Perceptions: The Enemy Within," in <http://www.apcss.org/Publications/Ocasional%20Papers/OPTaiwanThreat.pdf>.

9

SINGAPORE AND THE REVOLUTION IN MILITARY AFFAIRS

Tim Huxley

Singapore's military capability is, by most measures, the most advanced in Southeast Asia. The buildup of Singapore's armed forces and its national defense industry, as well as local defense research and development (R&D), reflects the determination of the People's Action Party (PAP) government to ensure the city-state's survival in a potentially hostile regional environment. However, these developments would not have been possible if Singapore had not possessed such a highly developed economy and well-educated population. These advantages, reinforced by increasingly intense interaction with the armed forces and defense industries of advanced industrial countries, have allowed Singapore to make substantial strides toward participation in the Revolution in Military Affairs (RMA) since the 1990s. Singapore has fielded increasingly sophisticated defense systems, particularly in the RMA-critical areas of precision weapons, command, control, communications, and computer-processing (C4), and intelligence, surveillance and reconnaissance (ISR). Integrated logistic support (ILS) is also well developed. Though the doctrinal and organizational innovation required to implement the RMA is so far rather less well advanced, Singapore seems to have made substantial progress toward establishing a relatively low cost "system of systems," which will far outclass the military capabilities of other Southeast Asian states for at least the next decade. And although Singapore's government has become increasingly concerned over asymmetric threats, it is clearly determined to maintain its emphasis on developing high-technology conventional military capabilities.

SINGAPORE'S DEFENSE POSTURE

Geopolitical circumstances have forced Singapore's government to take defense extremely seriously since the city-state separated from Malaysia in acrimonious circumstances in 1965. Though the government sees security holistically and has implemented a strategy known as Total Defence, which provides for the wholesale mobilization of the population and national resources in time of crisis or conflict, the military component of defense has always loomed large. Despite Singapore's small size and population, by the late 1990s, its armed forces were probably the best-equipped, best-trained and potentially most effective in Southeast Asia. The government routinely devotes 25–30 percent of its total annual spending (roughly 5 percent of GDP) to the armed forces. In 2003/04, Singapore's defense spending amounts to US$4.7 billion, by far the largest national defense effort in Southeast Asia.[1]

The development of the Singapore Armed Forces (SAF) has traditionally been based on the need to deter and, if deterrence fails, to defend against threats posed by its much larger neighbors to north and south, Malaysia and Indonesia. Secondarily, Singapore's military capability has provided a firm basis for security cooperation with friendly powers from outside the region. Preeminent among these extraregional associates is the United States, which developed into a quasi-ally during the 1990s. But Singapore has always based its defense on the potential necessity of defending itself without direct outside help.[2]

In developing their armed forces, Singapore's leaders have increasingly stressed the importance of exploiting technology to compensate for the city-state's lack of strategic depth and shortage of professional military manpower. Though the SAF can mobilize 350,000 personnel if necessary, only 20,000 of these are regulars: the remainder are conscripts and reservists. The SAF prizes its "technological edge," which has almost certainly provided it with conventional military advantages over any likely adversaries in its immediate region. In part, this technological edge has derived from purchases of advanced military equipment from overseas suppliers (e.g., F-16C/D fighter/strike aircraft from the United States during the 1990s), but it is also a product of Singapore's highly capable defense industry and defense R&D efforts.

Singapore's defense establishment clearly recognizes the RMA's significance. According to *Defending Singapore in the 21st Century* (DS21), MINDEF's most recent comprehensive defense policy statement (issued in February 2000):

> The revolution in military affairs will change the nature of warfare. Superior numbers in platforms . . . will become less of an advantage unless all these platforms can be integrated into a unified, flexible and effective fighting system using advanced information technologies. At the same time, the ever-increasing reliance on information technology means that protecting one's own information systems and disrupting the enemy's will become a major aspect of warfare.[3]

Placing the SAF's future development firmly in this new context, DS21 promised that the SAF would "exploit developments in the RMA, such as the integration of information technology (IT) into weapon systems" to achieve battlefield superiority.[4] As for Singapore's defense industry, "the digital battlefield of the future and the need for commercial technology in IT and communications will influence the approach we take to ensure that we sustain a technological edge."[5]

ECONOMIC AND EDUCATIONAL ADVANTAGES

In attempting to participate in the RMA, MINDEF and the SAF have been able to exploit important national advantages. Despite occasional setbacks, Singapore has achieved outstanding economic success since the 1960s. Rapid and sustained growth, deriving largely from the government's relentless efforts to direct the economy into higher value-added, more highly technological and more capital- and knowledge-intensive activities, gave Singapore Asia's highest and the world's fourth highest per capita GDP by 1997. Though Singapore was affected by the region-wide recession that began in 1997, it was not so badly hit as its neighbors: growth recovered from 1.5 percent in 1998 to 5.4 percent in 1999 and 10.1 percent in 2000. In 2001, the weakness of U.S. economic growth, world semiconductor sales, and other regional economies caused a collapse in manufacturing exports, causing an economic contraction of 2 percent. This was the worst recession since independence and although the economy recovered in 2002 with growth of 2.2 percent, the impact of the regional SARS epidemic on the tourism industry caused another downturn during 2003. Nevertheless, overall economic prosperity has underpinned high levels of defense spending, which have in turn allowed the SAF to benefit increasingly from high-technology equipment, modern infrastructure, and high-quality training facilities.

Singapore's sophisticated economy is increasingly based on the types of activity needed to support a local version of the RMA. By 1998, Singapore was ranked as "the world's fourth most information-driven economy" (after the United States, Sweden, and Finland).[6] The telecommunications, IT, and media industries were, by the late 1990s, the most important area for new investment in the economy's service sector. The government is pushing the economy increasingly toward R&D and in October 2000 unveiled its National Science and Technology Plan for 2001–05, under which it will invest US$4 billion in high-technology R&D with the aim of transforming Singapore into a "knowledge-based economy" (KBE). A highly developed IT and communications sector will be key to the KBE.[7]

Economic progress has both facilitated and necessitated dramatic advances in Singaporeans' educational achievement. The education system is now explicitly geared toward educating Singaporeans for the KBE. The government's 1997 education blueprint, *Thinking Schools, Learning Nation*, encompasses a key initiative aimed at "developing critical and creative thinking skills in the young" and a master plan for IT education,[8] the latter aimed at training teachers "to bring IT into schools to prepare future generations for the digital economy."[9] To help meet the government's target of increasing Singapore's trained workforce in the IT and communications sector from 100,000 in 2001 to 250,000 by 2010, the Institute of Systems Science aims to train 15,000 students annually by 2005.[10] In the mean time, information and communications technology already plays a prominent part in the lives of most Singaporeans. For example, under the first phase of the "E-government action plan," 130 key public services are delivered electronically, and are widely used. The government has also established a broadband infrastructure that can be accessed by all schools and more than 99 percent of homes.[11] Given that the majority of working-age Singaporean men, and many women, are involved in the SAF as regular, reservist or conscript personnel, this trend toward a better-educated population, increasingly familiar with information and communications technology, facilitates the SAF's absorption of the technologically sophisticated systems intrinsic to the RMA.

DEFENSE R&D

Singapore's strong orientation toward advanced technology is reflected in the vital role played by defense engineers and defense scientists from government agencies and the state-controlled defense industry, as well as "warfighters," in the Integrated Defense Development process, which guides the development of the SAF.[12] Though the republic's technological capacity is limited compared with that of larger industrial states, local development and upgrading of defense systems has provided access to military capabilities not available to Singapore through off-the-self purchase in the international market. In recent years MINDEF has increased its R&D budget substantially, from approximately 1 percent of defense spending in 1990 to 4 percent in 2000.[13] In real terms, this implied an increase from US$20 to 160 million. Working on the principle that "we cannot do everything ourselves,"[14] the republic's defense R&D establishment and industry have made strenuous efforts to acquire relevant technology through collaboration with both international and local partners. Crucially, Singapore's defense decision-makers realize that leads in military technology are ephemeral, and that defense R&D is a constant race to stay "ahead of the game."[15]

MINDEF claims that local R&D has provided the SAF with "silver bullets"—advanced systems that might prove militarily decisive in "extreme conditions."[16] But the essence of Singapore's RMA-relevant defense technology effort focuses on acquiring, developing, and integrating information and communications technologies for command and control with ISR systems and precision-guided weapons. The aim is to allow SAF combat units to locate, target, and destroy targets more effectively in the context of round-the-clock combined arms and joint-service operations. Developing and refining such a capability will be key to the SAF's continuing regional military superiority in the early twenty-first century.

The three "key pillars" underpinning Singapore's ability to harness technology for defense purposes, according to Deputy Prime Minister and Minister for Defense Tony Tan, are the SAF's highly educated personnel, a "versatile" local defense industry, and the Defence Science and Technology Agency (DSTA).[17] DSTA, a statutory body with a staff of 2,800 was established in April 2000 as the outcome of a major restructuring within MINDEF aimed principally at strengthening "technology acquisition and management."[18] The Agency is responsible for procuring equipment and services for the armed forces, for developing their infrastructure, and for managing defense R&D. DSTA's directorates of Air, Land, and Navy Material manage major equipment procurement; its Defense Information Systems directorate handles C4I programs.[19] DSTA collaborates extensively with local and foreign industry on:

- systems engineering and systems integration for new equipment;
- upgrading the performance and capability of existing systems;
- providing engineering support for selected weapon systems; and
- keeping abreast of relevant technology and advising the SAF on how to exploit it.

In addition, DSTA—through its CSO Development Laboratory—is the design authority for all MINDEF and SAF C4I systems' hardware and software, as well as for simulation systems, employing more than 400 engineers in divisions specializing in advanced technology, mission planning, and war gaming, air, land, naval, joint, communications, and dual-use (civil–military) systems. DSTA engineers' work on C4I uses "battle lab" modeling and computer simulations to produce "customized solutions" for Singapore's requirements.[20] The extremely active "dual-use solutions" branch has spearheaded the application of commercial-off-the-shelf (COTS) hardware and software within MINDEF and the SAF.[21] DSTA takes exploitation of dual-use technology seriously to the extent of funding small "start-up" companies conducting research relevant to SAF requirements. For example, with a view to improving battle simulation, DSTA might fund companies creating computer game software.[22]

Fundamental defense R&D are mainly the responsibility of the Defense Science Organization (DSO) which, with more than 800 engineers and scientists, is Singapore's largest R&D organization. DSO was originally concerned particularly with electronic warfare (EW), but its capabilities expanded considerably during the 1980s and 1990s.[23] In 1997, DSO was corporatized, becoming DSO National Laboratories, a nonprofit company affiliated with DSTA. This move was intended to improve DSO's efficiency by introducing more flexible, less bureaucratic commercial best practice in project management, subcontracting, technological alliances, intellectual property protection, and commercial ventures. In the personnel sphere, the intention was to equip DSO better to attract and retain the scientists and engineers who constitute its life-blood.[24]

In practical terms, limited resources dictate that while attempting to "stay close to the leading edge in the basic technologies,"[25] Singapore's defense R&D must target highly specific technological niches, chosen in close consultation with the SAF. These niches must offer substantial potential payoffs in terms of enhanced operational capability, lie within DSO's capabilities, and be assessed as worthwhile to investigate "in-house" for reasons of secrecy or because of the lack of alternative sources of the technology in question. DSO conducts R&D through 13 "centers of excellence" that work on areas of particular interest to MINDEF and the SAF. Those of particular relevance to the RMA include:[26]

- Advanced electronics and signal processing
- Decision support
- Information systems security
- Advanced systems (meaning guided systems)
- Communication systems
- Electronic warfare systems
- Radar systems
- Systems engineering
- Unmanned systems.

DSO's work on systems integration and software development has contributed importantly to supporting the SAF's RMA aspirations. Examples include the integration of radars, various types of surface-to-air missiles (SAM) and fighter aircraft with command and control elements into the air defense system during the 1980s, and the integration and capability optimization of new sensors and weapons (such as the Harpoon anti-ship missile and Barak SAM) for warships. R&D related to EW—which MINDEF has specifically identified as a "silver bullet"[27]—and to signals intelligence systems and information warfare are also key to the development of RMA capabilities.

DSO has played a central role in exploiting as well as generating "dual-use" technologies. In 1992, it inaugurated a "Technology Watch" program with the aim of identifying and monitoring "key emerging technologies for application in the SAF," including "ideas from technologically-advanced commercial sectors." DSO's corporatization in 1997 has facilitated cross-fertilization of R&D with local academic and research institutes, notably the National University of Singapore (NUS), Nanyang Technological University (NTU), the Institutes of Systems Science, Information Technology and High Performance Computing, as well as local companies, helping to expand further its exploitation of dual-use R&D.[28] Dual-use communications technology has been used in components and subsystems for radar systems, and COTS components in the DSO-developed Airborne Compute Engine, an ultra-fast military computer. Dual-use technologies have also played a key part in DSO's efforts to develop protection for the military communications and computing infrastructure against "information attack."[29]

Though DSO's focus remains on defense-related R&D on behalf of MINDEF and SAF, it "can no longer be assumed that defence R&D work will automatically be contracted to DSO."[30] When DSO was corporatized, MINDEF also established the Directorate of Research and Development (now part of DSTA) as its R&D "master planner" and buyer of R&D services for the ministry and the SAF—with the option of drawing on sources outside

DSO.[31] A potentially important new defense R&D source was established in September 2000 in the form of Temasek Laboratories, a collaborative research venture between DSTA and NUS, specializing in electromagnetics and aeronautics.[32]

DSTA and DSO have drawn extensively on foreign technological expertise in their R&D work, and are increasingly deeply involved in collaborative projects with foreign counterparts in areas of mutual interest, with the aim of maintaining the SAF's technological lead. Much of this cooperation is highly classified. This applies particularly to Singapore's defense R&D cooperation with Israel, but it is known that this has included work on electro-optics, training simulation, electronic warfare, anti-tank missiles, and unmanned aerial vehicles (UAVs).

During the 1990s, Singapore intensified its R&D collaboration with national defense science establishments in several other countries. Despite Washington's concerns during the 1980s that Singapore was a potentially untrustworthy end-user for high-technology military and dual-use exports,[33] U.S.–Singapore defense–technological cooperation has gathered pace in the new context of an increasingly close overall bilateral security relationship since the early 1990s. In 1999, Singapore joined the "demonstration phase" of the U.S.–U.K. Joint Strike Fighter (JSF) programme as a "Level 3 participant," allowing access to information regarding the program's technological progress.[34] In February 2003, the city-state increased its stake in JSF by joining the System Design and Development Phase as a "security cooperation participant" at a cost of US$50 million.[35] This may lead to substantial industrial cooperation and an order for the aircraft for Singapore's air force.

Collaboration with Sweden's Defense Research Establishment has been particularly close, and in 1997 Singapore and Sweden established a Joint Technology Development Fund to finance joint defense R&D projects.[36] Singapore has also explored possibilities for defense–technological collaboration with Australia (a 1993 Agreement for Cooperation in Defense Science and Technology led to joint projects in military communications), France (leading to establishment in 1997 of a Joint Technology Development Fund), South Africa, the United Kingdom, and Norway. In 1999, MINDEF set up a Defense Technology Office in Paris to coordinate Singapore's defense technology cooperation with France and other EU members.[37]

DEFENSE INDUSTRY

DSTA and DSO provide the crucial managerial and R&D underpinnings for Singapore's procurement of RMA-linked defense equipment. However, the role of the local defense industry, which is dominated by government-linked companies belonging to the Singapore Technologies (ST) group, has also been crucial.

Most of the ST companies were first established during the late 1960s and 1970s. Though the early focus was on producing unsophisticated arms and on maintenance, by the 1990s, the industry had developed strengths in retro-fitting and upgrading (particularly of combat aircraft), and the design, development, and production of artillery, armored vehicles, and small and medium-sized naval vessels. With DSTA as an intermediary, the industry has collaborated widely with foreign defense companies, which have facilitated its growing sophistication through Industrial Cooperation Programs involving technology transfer, agreed in connection with contracts for license production and "off-the-shelf" purchases for the SAF. Singapore's high-technology industrial base, sophisticated and competent defense R&D organization, and well-educated workforce have enabled it to absorb advanced defense-relevant technologies considerably more easily than other Southeast Asian states. By the end of the 1980s, Singapore's defense industry had become the most substantial, sophisticated, and diverse in Southeast Asia.

In 1997, four core ST businesses—ST Aerospace, ST Automotive, ST Shipbuilding & Engineering, and ST Electronic & Engineering—were grouped under ST Engineering (ST Engg), a new publicly listed company. The aim was to create the critical mass for a listed company, which would be more attractive to investors because of its size and smoother revenue stream, and which would benefit from cross-fertilization and rationalization of R&D operations. An additional subsidiary, ST Dynamics, was formed soon afterward to specialize in the development, production, and marketing of smart and guided weapons, and unmanned systems, areas that all involve close collaboration with Israeli industry.[38] In early 2000, ST Engg took control of ST's ammunition company, Chartered Industries of Singapore, which was joined with ST Automotive to form ST Kinetics, an integrated land systems arm. ST Engg now deploys a 11,500-strong workforce (including those employed by overseas subsidiaries) and the group's sales (60 percent of which were military) amounted to US$1.5 billion in 2002.[39] The government indirectly still holds 55 percent of the group's shares.

Although ST Engg has to compete against international suppliers, in effect the group remains a favored supplier to MINDEF and the SAF because of the leeway given during the procurement process to opt for local production or upgrading programs. Senior SAF officers frequently assume positions in ST Engg on leaving active military service, helping to ensure that the local defense industry remains closely attuned to the military's requirements.

ST Engg activities, which support the SAF fall into five main categories. At the most basic level, ST Engg continues to provide routine logistic support and depot-level maintenance. This support role expanded during the 1990s as the SAF—motivated by personnel shortages as well a desire for efficiency gains—commercialized more of its service, support, and logistic functions, to local industry's benefit.

ST Engg also continues to supply an extensive range of munitions for all branches of the SAF, and produces or assembles under license a wide range of weapons systems and other equipment, from Russian Igla man portable SAMs to French-designed frigates. In a fourth category of activity, ST Engg companies have—in close collaboration with DSTA and DSO—upgraded the operational capabilities of many SAF weapon systems (such as A-4S strike aircraft, AMX-13 light tanks, and missile gun boats) since the 1980s.

Finally, in conjunction with DSTA and DSO, ST Engg companies have developed and produced a range of new equipment for the SAF. It would be erroneous to call most of these systems "indigenous," as they have often relied heavily on imported design expertize. Over time, though, the systems produced have become increasingly sophisticated, reflecting local industry's expanding confidence and capabilities. As with the upgrade projects, close collaboration with DSTA and DSO has characterized these programs, which have included the SAR-21 rifle equipped with an integral laser-aiming device and the Bionix infantry fighting vehicle.

Behind the scenes, ST Engg—and particularly ST Electronics—is closely involved with DSTA and DSO in a range of highly classified C4 and ISR projects, usually only hinted at in official documents. Other key areas of RMA-relevant research involve developing computerized war gaming and simulation, information security systems, and offensive information warfare capabilities.[40]

ST Engg has recently been more open about some aspects of its RMA-relevant R&D. For example, the company has revealed its work on several UAV projects, including the Tailsitter ("smaller than a golf bag") and the Sparrow, a "palm-sized device." At the other end of the size range for UAVs, in conjunction with U.S. designer Burt Rutan, ST Engg and DSTA have drawn up plans since 1998 for a huge "battle management" drone (approximately the size of a Boeing 737 airliner), known as LALEE (Low-Altitude Long Enduring Endurance).[41]

At the tactical level, ST Electronics is working with DSTA on the Advanced Combat Man System, aimed at improving soldiers' situational awareness, "hitting power," and "battlefield survivability," especially in night combat. The system is based on a suite of advanced sighting and aiming devices linked to a backpack computer.[42] In another project looking to the future land battlefield, ST Engg has been involved with a U.S. company in developing a system for the U.S. Army's Future Combat System (FCS) requirement using unmanned robotic land warfare systems.[43] Together, these projects promise to revolutionize the technology available to Singapore's army.

THE SAF'S RMA-RELEVANT CAPABILITIES

Along with equipment procured in the international market, the plethora of RMA-relevant projects undertaken by the local defense R&D establishment and local industry have in many cases already fed through into capabilities providing a technological basis for Singapore's participation in the RMA.

Advanced Weapons Systems

Precision-guided weapon systems and their associated platforms have become ever more prominent elements of the SAF's capabilities over the last decade, increasingly providing the capability to hit targets more accurately and at greater range. Key activities for DSTA and ST Engg have been—and continue to be—the integration of guided weapons from diverse sources into the SAF's ships and aircraft and the adaptation and improvement of these missiles' performance. For this reason, access to software source codes is an important consideration for MINDEF when it purchases defense systems in the international market. U.S. restrictions on the transfer of such codes (e.g., in connection with the purchase of F-16C/Ds in 1993–4) have sometimes irked Singapore and have undoubtedly stimulated research aimed at overcoming this obstacle. Israel's willingness to supply Singapore with source codes is one reason for its success in marketing defense systems to the city-state.[44]

The navy fields large numbers of U.S.-supplied 90-km-range Harpoon long-range anti-ship missiles on its corvettes and missile gun boats in addition to the earlier, shorter range Israeli-made Gabriel weapons. By the end of the 1990s, the navy deployed a total of 120 anti-ship missile launchers, the largest number of any Southeast Asian navy. The new French-designed modified Lafayette class frigates, due to become operational from 2005, will be armed with the latest MM-40 Block II Exocet missiles. Since 1993, the navy has also armed most of its vessels with French Mistral/Simbad air defense missile systems. Corvettes are equipped with the Israeli-supplied Barak system, effective against anti-ship missiles as well as aircraft.[45] The Eurosam Aster air defense system is likely to be ordered for the new frigates.

Since the late 1990s, Singapore's air force has dramatically expanded its capabilities for air defense and for long-range precision strike with the acquisition from the United States of F-16C/D combat aircraft. By July 2001, no fewer than 62 of these aircraft had been ordered. Eighteen are already in service in Singapore, with 20 more due for delivery from late 2003, while others are based in the United States for training on a long-term basis. Singapore's Chief of Defense Force did not exaggerate when, in 1998, he claimed that procurement of these aircraft marked a "quantum leap" in the Singapore's air capability.[46] They are armed with AIM-7M air-to-air missiles, providing a beyond-visual-range engagement capability against hostile aircraft for the first time, and possibly Israeli Python 4 missiles. Although the U.S. government has allowed Singapore to purchase the AIM-120C advanced medium-range air-to-air missile (AMRAAM) for its F-16s, Washington requires—in order to help prevent a regional arms race—that they are stored in the United States and will only be delivered to Singapore in a crisis.[47]

In the strike role, the F-16C/Ds can—like earlier Republic of Singapore Air Force (RSAF) combat aircraft—carry Maverick air-to-ground missiles (both the AGM-65E, and the TV-guided AGM-65B for anti-ship strike), cluster bombs, and Paveway laser-guided bombs.[48] However, the new F-16s were also equipped with Sharpshooter navigation and targeting pods, enabling them to execute long-range precision-strike missions and to "self-designate" targets during day or night and in all weather conditions.

Singapore's F-16Ds are reported to have received special modifications on the production line, involving the installation of Israeli-supplied electronic countermeasures equipment. This may indicate that these two-seat aircraft are specially equipped for missions to suppress enemy air defenses.[49]

The RSAF's older fast combat aircraft—the locally upgraded F-5S in the air defense role and the A-4SU for strike—are approaching the end of their service life, and from 2006–07 will be replaced by a new multirole combat aircraft. A competition is presently underway to choose the new aircraft type, the Boeing F-IST, the Dassault Rafale and the Eurofighter Typhoon having been shortlisted in October 2003. This procurement program, which will involve substantial local industrial participation, is likely to lead to further substantial enhancement of the Singapore air force's air defense and long-range strike capabilities. For example, BAE Systems—which is marketing the Eurofighter—is offering Singapore an integrated package of systems virtually identical to those which are being purchased by the U.K.'s Royal Air Force, including the Meteor beyond-visual-range air-to-air missile, advanced short-range air-to-air missile (ASRAAM), together with Brimstone and Storm Shadow short-range attack and long-range cruise missiles.[50]

The army relies less than the SAF's other two branches on advanced technology solutions. Nevertheless, MINDEF has focused considerable efforts on improving land forces' mobility and firepower, particularly over the last decade. The most obvious accretion of firepower has come from the expansion of the 155-mm artillery inventory, particularly with the locally developed FH-88 and FH-2000 guns, more than 100 of which were in service by the late 1990s. Future artillery equipment will include self-propelled 155-mm guns and possibly also multiple rocket launch systems. Anti-armor capabilities have benefited from the infantry's adoption in 1990 of the 2-km-range Milan anti-tank missile, and since the late 1990s the 4-km-range Israeli Rafael NT-S Spike, a weapon described as a "computer with a warhead" and which MINDEF claims is effective even against explosive reactive armor.[51] Since the early 1990s, the air force has flown Fennec helicopters armed with TOW-2A missiles in the anti-armor role. This helicopter anti-armor capability is being upgraded dramatically with AH-64D Apaches equipped with highly sophisticated Longbow fire control radars, allowing them to designate their own targets for attack with the latest "fire and forget" version of the Hellfire laser-guided anti-tank missile.[52]

C4 and ISR Capabilities

Singapore's C4 and ISR capabilities are sophisticated and highly integrated, and increasingly provide the means for the effective operational coordination of the SAF's growing firepower. In 1991, MINDEF requested proposals for a Singapore-wide command, control, communications and intelligence network, based on microwave and fibre-optic channels and including links to air and maritime surveillance assets.[53] During the early 1990s, a branch of DSTA developed such a network, which is focused on an underground

Armed Forces Command Post.[54] According to MINDEF, a "computerised command and control system provides up-to-the-minute updates of the battlefield situation, including the disposition of friendly and hostile forces."[55] The Air Force Systems Brigade provides a full picture of the air situation, integrating data from ground-based radars and airborne early-warning and control aircraft, and is "charged with the operational and tactical control of all airborne aircraft."[56] The navy's Coastal Command contributes a central sea surveillance facility, using data from shore-based military and civilian radars, ships at sea, maritime patrol aircraft, and shore-based electronic and signals intelligence.[57]

MINDEF has invested heavily in cutting-edge IT for command and control purposes. As early as 1990, Brigadier-General Lee Hsien Loong, then second defense minister (services), claimed that IT could potentially provide the SAF with a "strategic edge" over an opponent. IT applications, which he foresaw included computer-based tools and staff aids to assist SAF commanders and staff officers in operational planning and decision-making.[58] COTS computer technology has been widely exploited: for example, Minister for Defense Tony Tan claimed in 1996 that Singapore was "among the leaders in the world in using [COTS] computers in real-time command and control systems," notably in integrating air defense sensors and weapons systems for the air force's Air Defense Systems Division.[59] And, in the intelligence sphere, MINDEF's Joint Intelligence Directorate collaborated closely during the 1990s with the Computer Systems Organization (now part of DSTA) to develop INSIGHT, a system enabling "efficient and effective means of information gathering, processing, retrieval as well as timely dissemination" in support of SAF operations.[60]

New technology is also impacting on tactical command and control. In the late 1990s, the army introduced the battalion-level Artillery Tactical Command and Control System (ATCCS) to compute and manage firing data more accurately and rapidly. In May 2001, Chief of Navy Rear-Admiral Lui Tuck Yew spoke of the "quantum improvement of our C4I networks," which combined with the use of muzzle velocity radars for improving gun-firing accuracy had—he claimed—allowed the navy "to move into the new domain of network warfare relying on precision strike and tactical force dominance."[61] Naval vessels have incorporated increasingly sophisticated combat information centers, with important elements developed by ST Electronics. The modified Lafayette-class frigates will be equipped with a C4I suite integrating ship management, weapons control and communications developed by ST Electronics and the navy under the Intelligent Naval Defence Platform 21 program.[62]

One important dimension to Singapore's emerging C4 and ISR capabilities is MINDEF's use of satellites for both communications and surveillance purposes. The republic's first communications satellite was a joint project between Singapore Telecom (SingTel) and Taiwan's Chungwa Telecom, and was launched from French Guiana in August 1998. This satellite, ST-1, was built by the Anglo-French Matra-Marconi company and is designed for both broadcasting and telecommunications (including data, telephony, and multimedia) purposes, with its 30 transponders shared equally between Singapore and Taiwan. Its "footprint" covers "the whole of Asia."[63] SingTel has been keen since the late 1990s to increase its satellite access and in January 2001 announced that it would lease 15 transponders on Apstar V, a Chinese-owned but U.S.-built satellite due for launch in 2003.[64]

Another satellite program began in 1995 when NTU signed an agreement with the United Kingdom's University of Surrey covering collaborative research and training in satellite engineering, communications satellites, and low earth orbit technology. As a result, in April 1999 a jointly designed 350-kg mini-satellite was launched from a Russian space base in Kazakhstan and has subsequently passed over Singapore every 90 minutes.[65] It carries NTU designed components in a package referred to as the Merlion Communications Payload, which has been used to research "real-time mobile communications and satellite linking of vehicles in GPS-based fleet tracking/control."[66] The next phase will involve X-SAT, a 100-kg micro-satellite designed and built at NTU in partnership with DSO and scheduled for launch aboard India's Polar Satellite Launch Vehicle by 2006.[67] This will be the first of a planned "equatorial belt" of Singaporean micro-satellites, ultimately providing Singapore (and potentially other users located close to the equator) with round-the-clock access to clearer and faster satellite communications.[68] DSO's involvement underlines the project's military significance. Satellites increasingly play an important role in intelligence collection. Since 1995, the Center for Remote Imaging, Sensing and Processing (CRISP) at NUS has routinely downloaded images from European, French, and Canadian satellites, including the French Spot-4, launched in March 1998.[69] CRISP's role in monitoring marine pollution and forest fires in the region has received wide publicity, but its defense intelligence role has not been acknowledged officially. However, the 1-m resolution of these images is sufficient to generate militarily useful information on the location of ships, armored vehicles, and aircraft.

Eventually, Singapore may rely to a large extent on its own satellites for collecting intelligence imagery. The NTU–University of Surrey mini-satellite's cameras were reportedly returning "spectacular" imagery within weeks of launch, and X-SAT will have a remote sensing as well as a communications role. However, a joint project with Israel (which has reportedly already provided satellite imagery on a commercial basis) may eventually provide Singapore with a greatly enhanced space-based surveillance capability. Under an agreement signed in June 2000, Singapore will reportedly fund further development of Israel's Ofeq series of satellites, advanced versions of which will eventually be operated by MINDEF.[70]

A key aspect of Singapore's effective information dominance over its region is its ability to collect signals intelligence (SIGINT) through land-based systems operated by army signals battalions, as well as other systems are deployed on aircraft and naval vessels. SingTel's radio receiving station at Yio Chu Kang probably also contributes data to the overall SIGINT picture.[71] According to Australian defense analyst Desmond Ball:

> Some of Singapore's SIGINT capabilities, especially the systems acquired from
> Israel over the last decade but increasingly also some designed and developed
> indigenously by [DTG] and Singapore Technologies, are among the most
> advanced in the world. Overall, it amounts to a sophisticated capability which pro-
> vides Singapore with strategic COMINT concerning its neighbours (Malaysia and
> Indonesia); HF DF/ocean surveillance information, including a very detailed pic-
> ture of the maritime traffic in the Straits and waters surrounding Singapore; a
> comprehensive picture of the electronic order of battle (EOB) of its neighbours;
> and the most advanced electronic warfare (EW) capability in Southeast Asia.[72]

The air force's E-2C Hawkeye airborne early warning (AEW) and control
aircraft also contribute significantly to Singapore's intelligence picture and
ability to manage the air and sea battle, and provide over-the-horizon targeting for the navy's Harpoon long-range anti-ship missiles. Though the E-2Cs
have been upgraded locally with a new mission control system,[73] MINDEF is
considering how to maintain and enhance its airborne battle management
capability in the future. While the most obvious solution would be to procure a radar-equipped small airliner or regional jet (such as the Boeing 737
with Northrop Grumman's MESA radar, as purchased by Australia), some in
DSTA and ST Engg have proposed adopting a distributed network employing the LALEE UAV as a platform performing a range of C4 and ISR functions.[74] Though evidently still at an early developmental stage, LALEE
might ultimately carry not only AEW radar, but also sensors akin to those
used for battlefield monitoring and stand-off radar reconnaissance by the
U.S. Joint Surveillance Target Attack Radar System.

Integrated Logistic Support and Maintenance

One of Singapore's great advantages in developing RMA-type capabilities is
the SAF's access to integrated logistic support and maintenance, which are
largely missing elsewhere in Southeast Asia. Crucially, the philosophy of Life
Cycle Maintenance (LCM) guides the procurement process, with the intention of ensuring the reliability and maintainability of the SAF's equipment,
reducing logistic support requirements and facilitating more effective use of
the procurement budget.[75] DSTA's target is that LCM costs should be no
more than 60 percent of the initial purchase costs of any particular procurement program.

The central role of local industry (essentially ST companies) in providing
logistic support for the SAF has increased substantially over the last decade,
particularly as a result of the commercialization of noncombat support services, such as the army's General Supply Base and Ordnance Supply Base,
depot-level air force maintenance, and parts of Naval Logistics Command.
Comprehensive logistic and maintenance support from the state-controlled
local industry is integral to the SAF's potential for sustained RMA-type operations and provides a key strategic advantage over other regional states.

Organizational and Doctrinal Issues

It is clear that, in purely technological terms, Singapore is acquiring many of the necessary prerequisites for participation in the RMA. Moreover, although MINDEF and the SAF have not so far implemented the doctrinal and organizational innovations necessary to absorb these technologies into an effective "system of systems," there are important indications that the SAF has begun laying the foundations for radical transformation.

Even before discussion of the RMA became voguish, SAF 2000, a planning blueprint adopted in 1988 as the result of a major force structure review, brought significant changes to organization and doctrine, particularly in the army. Under Army 2000, a single-service derivative of SAF 2000, army doctrine stressed offensive combined arms operations and the conduct of a "24-hour battle." In organizational terms, the most important change under Army 2000 was the introduction of genuine (as opposed to nominal) combined arms divisions, each including an armored brigade as well as two infantry brigades, even in peacetime. Another innovation was the establishment of a light rapid deployment division trained for air-mobile and amphibious operations. In the mid-1990s, the organizational evolution went a step further with the integration of reservist and active units within the three combined arms divisions.[76]

SAF 2000 also brought much greater emphasis on joint-service cooperation, and from 1994 established the Integrated Warfare concept as the basis for a doctrinal framework that attempted to integrate and exploit synergies in the three services' capabilities through a joint-service command and control system. Because of the SAF's relative youth as an organization, small regular cadre and the lack of strong single-service traditions, institutional obstacles to joint operations are considerably less than in the case of longer established national armed forces. As a result of this new emphasis on joint-service operations, in 1989 the air force established a Tactical Support Wing, which became Tactical Air Support Command (TASC) in 1991 with responsibility for planning, coordinating, and providing air support for the army and navy. One key TASC activity is operating UAVs in support of the army. The increasing emphasis on joint-service cooperation was also clear in the establishment in 1995 of a tri-service officer training academy, the SAFTI Military Institute. In addition, the Tri-Service Staff Course, which is conducted six times a year for a total of up to 240 officers, is aimed specifically at furthering the SAF's Integrated Warfare capability.[77] Joint-service exercises have been held routinely since the 1990s.

MINDEF's commitment to exploiting new information and communications technologies to give the SAF a "strategic edge" in the area of C4I was clear even in the late 1980s and early 1990s.[78] In 1992, it was reported that the SAF planned operations based on a "radio electronic combat" doctrine that integrated electronic warfare with reconnaissance, physical disruption, and deception.[79] However, this doctrinal emphasis increased greatly under Army 21, the planning blueprint that has guided development of the SAF's land component since April 1999. Army 21 was written in the context of the RMA and emphasizes the development of information capabilities, deriving from the "integration of command, control, communications and sensor systems," sufficient to achieve "dominant battle-field awareness."[80]

Senior MINDEF officials (from the defense minister downward) and ranking SAF commanders speak the language of the RMA with a high degree of fluency, and evidently recognize the military component of a broader problem with which Singapore's leaders have been grappling since the 1990s: how to encourage Singaporeans to be more creative in order to retain and enhance the city-state's competitive advantages. A key problem in relation to the RMA is that Singapore's military command and control have in the past tended to be rigid and strictly hierarchical, with effective authority concentrated at the higher levels of MINDEF and the SAF. A reluctance to delegate authority to middle-level and junior commanders has been characteristic. For example, air force squadron commanders have hitherto been able to exercise little operational initiative compared with their Australian or British counterparts. The SAF's lack of organizational flexibility has been reinforced by not only the political and administrative system, which has tended not to reward individualism or creativity, but also by the local cultural milieu in which respect for elders and seniors, and considerations of "face," have traditionally been central features.

As in other areas of competition it is evident that, in the field of defense, technological superiority alone is not sufficient for Singapore to come out on top. New information and communications technology has evidently stimulated much thinking within the SAF about the need for new command and control doctrines and new forms of military organization. In 1999, the Singapore air force's Chief of Staff, Brigadier-General "Rocky" Lim, pointed out that, by providing rapid access to more information, the latest IT applications increase the pressure for decision-making at lower levels in the chain of command. According to Lim, this "could change your entire doctrine of air warfare."[81] The influence of intensified interaction with Western armed forces, which already practice more decentralized command and control, may also push MINDEF and the SAF to delegate operational authority to lower levels of command more effectively. This applies most obviously in case of the air force's long-term training programs in the United States, Australia, and France, but all three services benefit from extensive interaction with Western forces that are themselves going through fundamental doctrinal and organizational change.

However, glimpses of internal debates within the armed forces, which are revealed in sources such as *Pointer* (the official SAF journal aimed at commissioned officers) suggest some impatience among younger middle-ranking officers for doctrinal and organizational change, which would lend greater substance to Singapore's incipient RMA. As early as 1992, one young army officer (the commander of a semi-elite army Guards battalion) argued that the SAF could gain an edge over opponents by adopting the German military philosophy of *Auftragstaktik*, involving considerable decentralization of command and control, and greater expectations of initiative on the part of lower-level commanders and even individual soldiers:

> Our Asian heritage has unfortunately...put too much premium on the value of "face." We are exceedingly hierarchy-conscious to the extent that constructive criticism is extremely rare from bottom-up. It will take much time and deliberate effort to dispel the fear of...subordinates to speak up if they think their superiors are in the wrong, and for the latter to accept constructive criticism.[82]

Writing almost a decade later, a more senior SAF staff officer returned to this theme, pointing to both the German army's *Auftragstaktik* and the Israeli army's similarly decentralized command system, both based heavily on the initiative of commanders and soldiers, as examples to be followed in implementing Army 21.[83]

More recently, several *Pointer* articles have argued for major organizational change within the SAF in response to technological developments. The essence of these arguments is that the SAF should adopt what one officer termed a "flatter and more network-based system."[84] More specifically, another officer has indicated that Army 21 may just "put new wine into old bottles," and argues in favor of "streamlined and flattened military organizations," which will "allow the SAF to compress the time needed for battle-procedure and decision-making" while at the same time reducing the vulnerability of the army to a preemptive enemy attack. Following the examples of the U.S. Army's Force XXI and the French brigade-based army, he proposes that the Singapore army's basic combined arms units should be organized around brigades rather than divisions.[85]

At the beginning of the present decade, Singapore's military establishment began to consider issues related to military transformation in greater depth. The most significant indications of the potential for radical change in Singapore's military thinking and organization came in early 2003 when MINDEF and the SAF established a Future Systems Directorate (FSD) commanded by a one-star officer known as the "Future Systems Architect." FSD, which has been allocated responsibility for managing 1 percent of the defense budget (approximately US$47 million in 2003–04), is charged with challenging established military thinking to enable the SAF to cope effectively with a rapidly changing and unpredictable strategic environment. FSD is complemented by the SAF's Center for Military Experimentation (CME), which will use sophisticated simulations in its "battle labs" to "develop and evaluate new war-fighting concepts by creating an environment for exploration, experimentation and demonstration."[86] The emphasis, at least initially, is on exploiting C4I systems as force multipliers.

SINGAPORE'S STRATEGIC FUTURE: HOW RELEVANT IS THE RMA?

The great challenge for MINDEF and the SAF in the future will be to develop new doctrines and organizational forms that enable exploitation of advanced technologies in ways that are relevant to the city-state's changing strategic circumstances. Singapore's regional security environment has deteriorated significantly since the economic recession of 1997–98 and there are few signs that the city-state's strategic circumstances will improve in the foreseeable future. Relations with Malaysia remain unstable and unpredictable, being driven in large part by economic, social, and political developments there that are outside Singapore's control. War between Singapore and Malaysia still remains a remote possibility, but its prospect has entered politicians' and media commentators' discourse on bilateral relations since early 2002.

Malaysian defense procurement plans mimic the SAF's efforts to develop RMA-style capabilities. The Malaysian armed forces' plans for the rest of the decade emphasize not only acquisition of advanced weapons such as Su-30 advanced multirole combat aircraft, long-range anti-tank missiles, airborne laser target designators, and medium-range air defense missiles, but also much greater use of information and communications technology and electronic warfare.[87] This suggests that maintaining the SAF's precious technological edge may become an increasingly expensive business for Singapore. However, taking into account Malaysia's educational and technological shortcomings, the lack of coherence in its procurement strategy, and Singapore's far superior defense–industrial and R&D capabilities, there seems little doubt that the SAF will be able to retain the upper hand for the foreseeable future.

In these uncertain circumstances, Singapore's leaders—while never pointing at any specific threats—have repeatedly emphasized the continuing importance of the republic's military instrument as a deterrent. The ability of Singapore's defense establishment to continue developing, integrating, and developing operational concepts for the advanced information and communications technologies extensively employed for command and control with satellite and other surveillance systems (including AEW, maritime patrol, and tactical reconnaissance aircraft, UAVs, and ground-based radars), and with precision-guided weapons, will be key to the SAF's continuing regional military superiority. The aim will be to allow the SAF (particularly the air force, navy, and artillery) to locate, target, and destroy targets more effectively in the context of round-the-clock combined arms and joint-service operations. At the same time, greater emphasis on criteria of range and endurance in selecting major platforms (principally ships, submarines, and aircraft) will provide Singapore with an artificial form of strategic depth by allowing the SAF to fight at greater distance from home.

However, like their counterparts in other states attempting to engage in the RMA, Singapore's security planners have needed to consider possible asymmetric counters to their probable conventional military superiority. Social and political developments in Indonesia pose a new type of security concern for Singapore. Continuing social, economic, and political instability, together with intensifying secessionist and intercommunal conflict around Indonesia's periphery, have raised the possibility of a "complex emergency" on Singapore's doorstep involving a breakdown in law and order, warlordism, communal conflict, piracy, hostage-taking, unregulated population movements, famine, rampant disease, and environmental catastrophe. Though unlikely, it is conceivable that the SAF could be drawn into diffuse, low-intensity operations around the periphery of Indonesia (in the Riau Islands to Singapore's south, for example) if the situation there deteriorated significantly.

Other new challenges—from either governments or nongovernmental groups—might include various combinations of bombings, the use of weapons of mass destruction (particularly chemical or biological agents), or information attacks, aimed at Singapore's civilian population and national infrastructure as well as military targets. Contamination of Singapore's water supply, for example, could be a particular effective asymmetric weapon. Though countering such asymmetric threats would largely be the responsibility of "Home Team" nonmilitary agencies under the Ministry of Home Affairs (principally the police and civil defense force), MINDEF claims that the SAF has a range of capabilities relevant to such contingencies (e.g., the army's Special Operations Force in the antiterrorist role). According to Deputy Prime Minister and then Minister for Defense Tony Tan, during 2000–01 MINDEF and the SAF, working with the "Home Team," "made good progress" in developing "concepts, frameworks and operational plans" in relation to potential low-intensity conflict.[88]

The September 11 attacks in the United States and the Singapore authorities' arrest in December 2001 of 15 members of Jemaah Islamiah, a terrorist group allegedly linked to al-Qa'ida, in connection with a plot to attack local targets accentuated the city-state's concerns over potential asymmetric threats. The main impact on Singapore's security and defense planning was to reinforce the validity of the long-established idea of Total Defense, which involves nonmilitary agencies as well as MINDEF and the SAF in insuring Singapore's security.[89] In November 2001, the government announced that it would implement a "homeland security" strategy involving closer cooperation between MINDEF and the home affairs ministry, and the SAF and police.[90] By January 2002, a National Security Secretariat had been set up to oversee this cooperation against "non-conventional threats and new security challenges," and the government emphasized that it would need to allocate greater resources to the SAF, police, and civil defense force to help them counter new threats.[91]

Particularly in light of recent events, it is clear that Singapore's developing RMA-type capabilities do not provide a panacea for its security needs. However, they are not necessarily irrelevant to low-intensity concerns. For example, the greatly improved ISR capabilities likely to be generated by Singapore's investment in UAVs and satellites will be highly relevant to the monitoring of population and shipping movements to the south. Moreover, technological improvements in the capability of ordinary infantry soldiers, ranging from the SAR-21 rifle to the Advanced Combat Man System, have a wider utility than simply on a high-intensity battlefield against a conventional enemy. Information security systems may be as useful in protecting "critical national infrastructure" (such as public utilities and air traffic control) against "cyber-terrorism" as they are in defending military C4I systems against attacks by opposing armed forces. New thinking about military doctrine and organization may produce more flexible armed forces able to adapt quickly to a rapidly evolving spectrum of threats.

Conclusion

Singapore's resources for defense R&D and even for military procurement are slim compared with those available to the major Western military powers. To put Singapore's defense budget in perspective: in approximate terms it amounts to less than 2 percent of the United States' or 12 percent of Japan's military spending. Moreover, even the United States' close military allies in Europe, such as the United Kingdom (which spends more than seven times as much as Singapore on defense) face considerable difficulties in keeping up with U.S. technological advances and insuring interoperability. Realistically, in relation to the major RMA players, Singapore's incipient advanced technology military capabilities might best be described as "RMA lite."[92] However, Singapore is not expanding the SAF's technological capacity primarily with a view to participating in U.S.-led coalition warfare. Interconnectivity with U.S. forces may be a welcome spinoff, but MINDEF's principal aim is to develop forces that in the last resort can defend Singapore against regional threats autonomously.

For less than US$5 billion annually, MINDEF and the SAF provide Singapore with a remarkable range of military capabilities. In Singapore's immediate regional context, these capabilities presently far outclass those of any potential opponent in conventional military terms. Singapore possesses highly educated and IT-literate military, research and industrial personnel, and its defense-industrial and R&D establishment has set up an extensive network of international links. For these reasons, it can almost certainly sustain its conventional military advantage for at least the next decade. Indeed, if Singapore develops military doctrine and organization, that allow the SAF to exploit its C4, ISR and firepower to the full, it may be able to assure continued military superiority in Southeast Asia even if potential regional adversaries (notably Malaysia) are eventually able to catch up in technological terms. The greater challenge, however, will be to adopt the RMA to the city-state's evolving strategic environment, in which asymmetric threats have since the late 1990s unexpectedly become prominent.

Notes

1. "Expenditure Overview: Ministry of Defence," Singapore Ministry of Finance Website, www.mof.gov.sg/budget/budget_2003/exp_overview/budget_mindef.html
2. For a fuller assessment of Singapore's defense policy, threat perceptions, and strategy, see Tim Huxley, *Defending the Lion City: The Armed Forces of Singapore* (St. Leonards, NSW, Australia: Allen & Unwin, 2000) pp. 24–72.

3. *Defending Singapore in the 21st Century* (Singapore: Ministry of Defence, 2000), p. 10.
4. Ibid., p. 75.
5. Ibid., p. 69.
6. *Singapore 2000* (Singapore: Ministry of Information and the Arts, 2000) pp. 125–26.
7. Natalie Soh, "$7b thrust for Science and Tech," *Straits Times Weekly Edition*, October 28, 2000, p. 1.
8. *Singapore 2000*, p. 215.
9. "E-government action plan," *Financial Times* (London), December 6, 2000.
10. Chang Ai-Lien, "Boost for Infocomms Lessons go Online," *Straits Times Weekly Edition*, July 7, 2001., p. 1.
11. "E-government Action Plan."
12. Keynote address by Dr. Tony Tan, Deputy Prime Minister and Minister for Defense, at the Signing Ceremony of the Statement of Intent for the Temasek Defence Systems Institute, Singapore Government Press Release, July 11, 2001.
13. Keynote address by Dr. Tony Tan at the launch of Temasek Laboratories, MINDEF Internet Webservice, September 6, 2000.
14. Keynote address by Dr. Tony Tan at the Signing of the Statement of Intent for the Temasek Defence Systems Institute, MINDEF Internet Webservice, July 11, 2001.
15. Speech by Deputy Prime Minister Brigadier-General (NS) Lee Hsien Loong, MINDEF Internet Webservice, October 3, 1997.
16. Speech by David Lim, Minister of State for Defence and information, MINDEF Internet Webservice, October 29, 1999.
17. Speech by Dr. Tony Tan at the Launch of the Defence Science and Technology Agency, MINDEF Internet Webservice, March 29, 2000.
18. Keynote address by Dr. Tony Tan at the launch of Temasek Laboratories, MINDEF Internet Webservice, September 6, 2000.
19. For a description of Singapore's defense procurement process, see Huxley, pp. 175–177.
20. David Boey, "Singapore's New Drones make Public Debut," *Business Times* (Singapore), February 26, 2002.
21. Prasun K. Sengupta, "Investments in defence R&D pay off," *Asian Defence Journal*, July 1999, p. 26.
22. Chan Kay Min, "Defence Agency to Fund Start-ups," *Straits Times*, February 14, 2002.
23. "Defence Science Organization. Defence R&D at its Best," *Pioneer*, November 1989, p. 18; Speech by Deputy Prime Minister Brigadier-General (NS) Lee Hsien Loong, MINDEF Internet Webservice, October 3, 1997.
24. Ibid.
25. Speech by Dr. Tony Tan, MINDEF Internet Webservice, November 4, 1998.
26. DSO National Laboratories website: www.dso.org.sg/
27. *Defending Singapore in the 21st Century*, p. 63.
28. DSO National Laboratories website, www.dso.org.sg/tech-colloboration.html [*sic*]
29. "Communication Technology—the Vital Link in Warfare," *Pioneer*, March 1999, p. 13; "Corporatisation of DSO," Media Releases, MINDEF Internet Webservice, March 14, 1997; *Defending Singapore in the 21st century*, pp. 66–67.

30. "DSO, as a National Resource, is Uniquely Suited to Build up Technologies for both Military and Commercial Applications," MINDEF Internet Webservice, October 3, 1997.
31. *The MINDEF/SAF Fact Book* (Singapore: Ministry of Defence, Ca. 1999) p. 21.
32. "Technology to Sharpen SAF's Edge," *Straits Times*, September 7, 2000.
33. "US and Singapore in Talks to Block Hi-tech Leakages," *Financial Times*, August 16, 1985; "RSAF Plans to update Skyhawks Blocked by Pentagon," *Business Times* (Singapore), March 27, 1986.
34. "Singapore Joins JSF, Australia Stays out," *Defense News*, May 10, 1999; "Singapore Signs Letter of Intent for Joint Strike Fighter Programme," MINDEF News Release, February 22, 2003.
35. Andrew Doyle, "Sharper focus," *Flight International*, February 19, 2002, p. 59.
36. "Singapore, Sweden set up US$5 million research fund," *Pioneer*, November 1997, p. 10.
37. "Dr Tony Tan visits France," MINDEF Internet Webservice, June 12, 1999.
38. "Newcomer with a Bulging Order Book," Singapore Survey, *Financial Times* (London), March 31, 1998; Prasun K. Sengupta, " 'We Have an Integrated Capability'," *Asian Defence Journal*, July 1999, pp. 22–25.
39. "ST Engineering: An Emerging World Class Player in the International Defence Market," *Military Technology* (6/2000) pp. 139–40; "The Focus" (annual report 2001), Singapore Technologies Engineering website, www.stengg.com/AR2001/index.html; "The Art of Engineering" (annual report 2002), Singapore Technologies Engineering website, www.stengg.com/AR2002/art. html
40. On information warfare, see Damon Bristow, "Asia: grasping information warfare?," *Jane's Intelligence Review*, December 2000, p. 36.
41. Andrew Doyle, "Singapore Recruits Rutan to Work on Long Endurance UAV," *Flight International*, May 15, 2001, p. 5. "Low-altitude" could be misleading, in that the drone is intended to fly at 60,000 feet—"low" only in relation to orbiting satellites. "Long enduring" refers to the platform's expected service life.
42. David Boey, "ST Elec in Project to Hone Soldiers' Hitting Power," *Business Times* (Singapore), February 26, 2002.
43. Denesh Divvanathan, "ST Engg Plans Foray into China, South America," *Straits Times*, March 9, 2002.
44. "RSAF Turns to Israel for EW," *Jane's Defence Weekly*, October 10, 1992, p. 5.
45. "Navy buys Barak Anti-missile System," *Straits Times*, April 23, 1996.
46. Speech by General Bey Soo Khiang, Fort Worth, Texas, MINDEF Internet Webservice, April 9, 1998.
47. "Singapore Advances Fighter Procurement," *Jane's Defence Weekly*, March 6, 2002, p. 14.
48. "F-16C/Ds boost RSAF's fighting capability," MINDEF Internet Webservice, August 14, 1998.
49. "Singaporean F-16D Block 52s Reveal Israeli Design Heritage," *Flight International*, April 22, 1998.
50. John Fricker, "Singapore closest to finalizing future fighter preference," *Aviation Week ShowNews online*, http://www.awgnet.com/shownews/02asia1/airfrm08.htm
51. "Tank busters," *Pioneer*, October 1993, pp. 1–4; "Fact Sheet—Spike Anti-tank Guided Missile [ATGM]," MINDEF Internet Webservice, July 13, 1999.
52. Andrew Doyle, "Sharper Focus," *Flight International*, February 19, 2002, p. 60; "Singapore Inaugurates Apache Unit," *Defense News*, April 21, 2003.

53. *Asian Defence Journal*, May 1991, p. 76.
54. David Boey, "Defending Singapore: A Fragile City-state's Approach to Defence and Security," MA dissertation, University of Hull, 1996, p. 48.
55. *Defence of Singapore 1994–95* (Singapore: Ministry of Defence, 1994) p. 60.
56. "Enhanced National Air Defence Capability," MINDEF Internet Webservice, July 13, 1998.
57. David Boey, "Singapore's fleet gets boost from Navy 2000," *International Defense Review* (12/1995) pp. 67–68; "Regional maritime air power evolves," *Asia-Pacific Defence Reporter*, February–March 1999, p. 19.
58. "SAF to Widen use of Info Technology," *Straits Times*, January 20, 1990.
59. Speech by Dr. Tony Tan Keng Yam, MINDEF Internet Webservice, October 4, 1996.
60. Ibid., November 3, 1997.
61. Speech by Chief of Navy RADM Lui Tuck Yew at the Naval Platform Technology Seminar 2001, MINDEF Internet Webservice, May 10, 2001.
62. "Naval Shipbuilding Programmes Asia and the Middle East," *Naval Forces* (1/2000) p. 48; "Singapore Orders La Fayette Frigates," *Asia-Pacific Defence Reporter*, April–May 2000, p. 39.
63. "Singapore's First Satellite to Launch in May," *Singapore Bulletin*, March 1998, p. 17; "S'pore's first satellite blasts off into space," *Straits Times Weekly Edition*, August 29, 1998, p. 24.
64. "China to Launch APSTAR V in 2003," Xinhua news agency, January 8, 2001; "Loral and APT Satellite Agree to Joint Ownership of APSTAR-V Satellite," Loral website, http://www.loral.com/inthenews/020923.html
65. "NTU Launches First Satellite Successfully," *Straits Times*, April 22, 1999.
66. *Jane's Space Directory 2001–2002* (Coulsdon, Surrey: Jane's Information Group, 2001) p. 71.
67. Radhakrishna Rao, "Delhi's Commercial Space Ambitions Lifted as Nanyang becomes Fifth Overseas Client," *Flight International*, February 15, 2003.
68. "Plans for Home-made Micro-satellites," *Straits Times Weekly Edition*, April 11, 1998; Paula McCoy, "Work on First Made-in-Singapore Satellite to Begin," *Straits Times*, December 12, 2001.
69. "Crisp pictures from S'pore's eye in the sky," *Straits Times*, May 15, 1997; "New Eye in the Sky for Close Look at Region," ibid., March 25, 1998.
70. "Israel, Singapore to Sign Satellite Deal," *Jane's Defence Weekly*, July 5, 2000, p. 2.
71. Desmond Ball, "Signals intelligence (SIGINT) in Singapore," unpublished paper, 1995, pp. 19–25.
72. Ibid., p. 30.
73. "Eyes and Ears of Air Force Upgraded," *Straits Times*, April 24, 2001.
74. David Boey, "Development of LALEE Drone Started 3 Years Ago, says Mindef's Chief Scientist," *Business Times* (Singapore), May 11, 2001; David Boey, "Singapore's New Drones Make Public Debut," ibid., February 26, 2002.
75. "Applying Life Cycle Management," *Pioneer*, December 1991, p. 35.
76. For details see Huxley, pp. 123–126.
77. "Officers from the Army, Navy and Air Force to Train Together," *Straits Times*, October 7, 1998.
78. "Information Technology. Giving the SAF a strategic edge," *Pioneer*, March 1990, pp. 14–17.
79. Prasun K. Sengupta, "Singapore and the Army 2000 Plan," *Military Technology* (7/1992) p. 73.

80. "Building the 21st Century Warrior—Army 21," *Pioneer*, May 1999, p. 13; *Defending Singapore in the 21st Century*, p. 30.
81. "Millennium force," *Flight International*, June 16, 1999, p. 67.
82. Major Peter Gwee Chon Lin, "Auftragstaktik. A Philosophy for Management, Training and War," *Pointer*, 18: 4 (October–December 1992) p. 34.
83. Lieutenant Colonel Tan Kim Seng, "Initiative as the Fighting Power in the Army 21's vision [*sic*]", *Pointer*, 27 : 3 (July–September 2001).
84. Major Seet Pi Shen, "The Revolution in Military Affairs (RMA): Challenge to Existing Military Paradigms and its Impact on the Singapore Armed Forces," *Pointer*, 27: 2 (April–June 2001) p. 16.
85. Captain Fong Kum Kuen, "A Quantum Leap towards Knowledge Warfare: Revolution in Military Organizations in the SAF," *Pointer*, 27: 2 (April–June 2001) pp. 80; 92; 94.
86. David Boey, "Battle Lab to Help Reshape SAF War Muscles," *Straits Times*, July 12, 2003.
87. "Electronic Warfare to get Higher Priority in Defence," *Straits Times*, September 25, 2002.
88. Statement by Dr. Tony Tan at the Committee of Supply debate, March 8, 2001, Singapore Government Press Release, March 8, 2001.
89. "Sept 11 proves need for Total Defence, says DPM Tan," *Straits Times*, October 27, 2001.
90. Lydia Lim, "S'pore to have 'Homeland Security'," *Straits Times*, November 5, 2001.
91. Lydia Lim, "National Security Secretariat set up at Mindef," *Straits Times*, January 7, 2002; Dr. Tony Tan, Ministry of Defense Addendum to the President's Address, Ministry of Defense Press Release, March 31, 2002.
92. I am indebted to Malcolm Davis for this term.

10

Conclusion: The Diffusion of the Emerging Revolution in Military Affairs in Asia: A Preliminary Assessment

Thomas G. Mahnken

This volume has examined the efforts of five Asian militaries—those of Australia, China, Japan, Taiwan, and Singapore—to exploit the information revolution. Two are U.S. allies, two are friends, and one is a competitor and potential adversary. Influential groups within each country's military favor harnessing the information revolution to overcome existing problems and gain new benefits. These five cases nonetheless display considerable variation in terms of both the extent and nature of their exploitation of the emerging revolution in military affairs (RMA). While Asian militaries have paid close attention to U.S. debates regarding the emerging RMA and frequently adopt U.S. terminology in their own discussions of future warfare, none has sought to import U.S. methods wholesale. Rather, each has begun to develop its own, unique approach to information-age warfare. Each also faces considerable barriers to exploiting the emerging RMA. As a result, any assessment of the information revolution in Asia should consider alternative scenarios for the diffusion of the emerging RMA.

This chapter describes the paths along which information-age warfare methods are spreading within Asia. It also summarizes the pattern of innovation in the five militaries studied and examines the drivers of and barriers to pursuing the RMA among the five militaries.

PATHS OF DIFFUSION

The United States currently sits at the apex of the international military hierarchy. U.S. armed forces are in many ways the very model of a modern military. They have taken the lead in fielding information, stealth, and precision technologies and have spent considerable effort developing innovative doctrine and organizations. They have demonstrated the ability to strike globally with precision at night and in all weather. In Afghanistan in 2001 and

Iraq in 2003, U.S. armed forces demonstrated the ability to overthrow less capable adversaries while suffering negligible to nonexistent casualties. In addition, there is a widespread perception—accurate or inaccurate—that U.S. armed forces have employed information warfare (IW) extensively.

Despite such impressive capabilities, and despite the attention the U.S. military has accorded to developing new ways of war, a concrete U.S. "model" of a "post-RMA military" has yet to emerge.[1] While American defense analysts frequently speak in terms of "network-centric warfare" and "effects-based operations," for example, such concepts remain amorphous. Even within the Defense Department there is considerable debate over what weapons, concepts, and organizations are truly "transformational." The U.S. armed forces have yet to establish an information-age equivalent of the *Blitzkrieg* or—more to the point—the *Panzer* division: a well-defined set of concepts and organizations tailored to waging war in the information age. As a result, no single model exists for other states to adopt or counter.[2] While U.S. thinking regarding the RMA exerts leverage, it is as ideas to be studied rather than techniques to be copied.[3] Absent a battle-tested model, variation and asymmetry among militaries in the region will grow over time.

Influential groups in each of the militaries examined here believe that the growth and diffusion of information technology (IT) will bring about an RMA. Each of the states has expressed considerable interest in U.S. approaches to information-age warfare. Rhetorically, each military uses the term "revolution in military affairs" even if, as noted later, it does not view the U.S. approach to exploiting the emerging RMA as being wholly applicable to its circumstances. It is worth noting, however, that a consensus in favor of pursuing the RMA exists in none of these militaries. Rather, in each there are various schools of thought about force modernization and future warfare. RMA advocates are quite often in the minority, even if in some cases they enjoy high-level support.

In a number of cases, Asian militaries have imported U.S. terminology in an effort to identify new approaches to combat. Perhaps the best example of this is the term "information warfare," which while used widely connotes many different things to many different people. It is not, however, a lone instance. As James Mulvenon notes in his chapter, Taiwan's military frequently uses such terms as "precision strike," "dominant maneuver," and "information warfare" in discussions of future warfare. Similarly, the Australian Defense Force's (ADF) concept of future warfare emphasizes "effects-based operations" (EBO) and "network-centric warfare."[4] However, such rhetorical similarity often masks important conceptual differences. The ADF's concept of EBO, for example, is much more comprehensive than that espoused by the U.S. Navy and Air Force, involving the use of all elements of national power.[5] Assessments of foreign military concepts must delve beneath such surface-level similarities to expose differences in the meaning of these terms to different defense establishments.

U.S. concepts of information-age warfare have spread through several channels. First, Asian armed forces have paid close attention to American books and articles about the emerging RMA. Foreign writings about future warfare frequently lean heavily upon U.S. sources. Articles by defense analysts such as Eliot Cohen, Andrew Krepinevich, and William Owens feature prominently in foreign writings about future warfare. Early conferences on the RMA featured a veritable who's who of American RMA advocates from government, the defense industry, and academia.[6] Much as Basil Liddell-Hart and J.F.C. Fuller influenced how armies across the globe approached armored warfare during the 1920s and 1930s, American RMA advocates are shaping how other militaries approach information-age warfare.[7] The same is true of official U.S. documents on future warfare. Indeed, one suspects that official pronouncements such as *Joint Vision 2010* and *Joint Vision 2020* received more attention outside the U.S. military than inside it.

A second way that U.S. concepts have spread is through military-to-military contacts. U.S. allies such as Australia and Japan maintain close ties to the U.S. armed forces, and these provide an important channel for them to learn about U.S. activities. The Australian government has followed closely U.S. speculation and experimentation regarding emerging warfare areas. The Australian Department of Defense has maintained a close relationship with the Pentagon's Office of Net Assessment. In addition, the ADF has a liaison officer at U.S. Joint Forces Command in Norfolk, Virginia. The Australian Army followed the U.S. Army's "Army After Next" program quite closely.[8] The Royal Australian Navy has participated extensively in the U.S. Navy's Global War Game. Australian officers routinely participate in U.S. field exercises as well. Japan similarly maintains close ties to the U.S. armed forces. Japan Self-Defense Force (JSDF) officers attend U.S. professional military educational institutions. The Japan Maritime Self-Defense Force (JMSDF) has a liaison at the Navy Warfare Development Command in Newport, Rhode Island. Japan also participates in a range of bilateral and multilateral exercises and war games.

Those states that lack a formal alliance with the United States have had less of an opportunity to glean insights from military-to-military exchanges. For example, Arthur Ding argues that Taiwan's diplomatic isolation has limited its exposure to Western militaries. Still, Taiwan maintains a range of contacts with U.S. armed forces, including sending officers to attend U.S. war colleges. James Mulvenon concludes that military-to-military ties have in fact played an important role in shaping Taiwanese views of the RMA.

While the United States today offers the most prominent source of thinking about information-age warfare, it is not the only one. It is worth remembering that the U.S. government first became aware of the concept of an RMA through its analysis of Soviet military writings. Indeed, it was Soviet military scientists who pioneered the concept in the 1950s in an attempt to come to grips with the dramatic changes in the conduct of war that had come about due to the advent of nuclear weapons and ballistic missiles. And it was Soviet defense analysts who first argued that the information revolution was bringing about a new RMA. Over time Soviet, then Russian, thinking regarding future warfare became quite advanced.

Other militaries, such as those of China and India, have paid close attention to Soviet writings on future warfare. You Ji notes, Chinese military scholars first encountered the concept of the RMA through their contact with Soviet military thinking. Soviet and Russian writings regarding information-age warfare have influenced the Indian military as well.[9]

Each of the militaries examined in this volume has paid close attention to the battlefield performance of U.S. forces in Kuwait, Bosnia, Kosovo, Afghanistan, and Iraq. U.S. allies have had the advantage of participating in these conflicts. Australian forces fought side-by-side with American troops in the Gulf War, Afghanistan, and Iraq, while the JMSDF deployed forces to the Persian Gulf in support of Operation *Enduring Freedom*. This experience has doubtless given them considerable insight into U.S. capabilities. Nor are they alone. The Chinese military, for example, has studied the U.S. military's performance in recent wars extensively.[10]

Patterns of Diffusion

As Emily Goldman discussed in the first chapter of this volume, military organizations develop new approaches to combat, or transform, in several phases. The dynamics of transformation cannot be understood without also considering the diffusion, or strategic interaction dimensions, of military change, in large part because, as these case studies demonstrate, militaries adapt based on information about and interaction with other militaries. The first step in understanding the development of the RMA in Asia is to identify a set of indicators that enable one to take a snapshot of the status of innovation there. Based upon insights derived from the authors of our case studies, we amended the indicators of innovation proposed in the introduction to this volume, adding to the list leadership consensus in favor of new warfare methods, and the allocation of resources to support them. Table 10.1 summarizes our assessment of innovation in each of these militaries.

Australia

To date, Australian authors have conducted a number of studies of the future security environment and emerging warfare areas. In April 1999, the Australian government established the Office of the RMA within the Department of Defense to review the development of advanced technology and develop a strategy to extract the maximum value of the RMA for the ADF. The department subsequently promulgated a draft discussion paper on the RMA and the ADF. Each of the services formed a dedicated futures directorate. They have conducted experiments to explore new technology, doctrine, and organizations. They have also formed several units dedicated to new mission areas, most notably an information operations squadron in the Royal Australian Air Force.

While Australia has entered into a protracted and detailed dialogue with the United States regarding future warfare concepts, this process has yielded

Table 10.1   Indicators of innovation in Asia

	Australia	Japan	China	Taiwan	Singapore
Speculation					
Publications describing potential new combat methods	X	X	X	X	X
Establishment of official organizations to study recent wars	X	X	X	X	X
Study of foreign innovation efforts	X	X	X	X	X
Experimentation					
Establishment of organization(s) charged with experimentation	X			X	X
Formation of experimental military units	X		?	X	?
Experiments with new warfare methods	X		X	X	X
War gaming of new warfare methods	X		X	X	X
Implementation					
Leadership consensus in favor of new warfare methods	X		X		X
Allocation of resources to support new warfare methods	X		X		X
Development of formal transformation strategy	X		?		X
Establishment of innovative military units	X		X	X	
Revision of doctrine to accommodate new ways of war					
New branches, career paths			X		
Field training exercises with new doctrine, organizations					

an Australian approach to information-age warfare that differs markedly from that of the United States. While the Australian Department of Defense initially adopted the term "Revolution in Military Affairs," it soon dropped it in favor of the term "Knowledge Edge" as a means of emphasizing Australia's unique approach. As the 1997 publication *Australia's Strategic Policy* defines it, the Knowledge Edge is "the effective exploitation of information technologies to allow us to use our relatively small force to maximum effectiveness."[11] Australian strategic analysts believe that the American method of exploiting the RMA is poorly suited to Australia's circumstances. In particular, they see the U.S. approach as lying far beyond Australia's limited fiscal and technological means.

The ADF has emphasized advanced intelligence, surveillance, and reconnaissance capabilities. The Defense Science and Technology Organization has established the *Takari* program to ensure that the ADF has an integrated C3I and information operations capability. It has created the position of Chief Knowledge Officer to manage these assets. The first occupant of the position, Air Vice Marshal Peter Nicholson, was an early proponent of the RMA in Australia.

On the other hand, the mere establishment of organizations does not guarantee that they will innovate. It is unclear, for example, how much influence the Chief Knowledge Officer has over the services, just as it is unclear how much pull the futures directorates have within each service. Moreover, the development of innovative doctrine within the ADF is still in an early stage. The Army is exploring concepts for inserting forces in the face of an anti-access threat under the rubric of Maneuver Operations in a Littoral Environment, or MOLE. The services are interested in information operations and new command-and-control concepts. However, the future of these efforts remains uncertain.

While Australia purchases many of its major weapon systems from the United States, the ADF has also displayed a great deal of technological innovation. One niche that Australia has pursued vigorously is over-the-horizon radar (OTHR). The Jindalee Over-the-Horizon Radar Network (JORN) was developed to provide a security shield for Australia's remote northern approaches. Designed to monitor air and sea movements across 37,000 km of unprotected coastline and 9 million km^2 of ocean, the network has the ability to track ships and aircraft (including stealthy ones).[12]

In at least one area, the United States has emulated an Australian innovation. In 1999, the ADF leased a high-speed ferry, rechristened the HMAS *Jervis Bay*, to support the deployment and sustainment of forces in East Timor. Inspired by the vessel's performance, the United States leased a fast catamaran—the USS *Joint Venture* experimental high-speed vessel (HSV-X1)—from INCAT of Tasmania to explore new concepts for amphibious and special operations missions.[13] Indeed, U.S. Navy SEALs used the vessel extensively during the Iraq War. The Marine Corps has leased another Australian fast ferry—the Austal TSV 101—for experimental purposes.

Japan

Japan's limited efforts at transformation demonstrate that access to IT is a necessary but insufficient ingredient for exploiting the RMA. China's growing power, the prospect of a conflict across the Taiwan Strait, the potential for instability on the Korean Peninsula, and uncertainty over the future role of the United States in Asia have all prodded Japan into reexamining its security requirements. Following North Korea's 1998 *Taepo-Dong I* missile test, Tokyo decided to develop its own reconnaissance satellite constellation. It has also begun to cooperate more closely with the United States on theater missile defense.[14] Moreover, the rise to power of the Koizumi government and the global war on terrorism have made Japan more assertive. The deployment in late 2001 of JMSDF warships to support Operation *Enduring Freedom* in the Persian Gulf and the deployment in 2004 of JGSDF troops to support Operation *Iraqs Freedom* in Iraq represent significant firsts.

Japan has also expressed interest in a more radical transformation of its defense posture. In December 2000, the Japan Defense Agency (JDA) released a study paper examining the implications of the information revolution for Japan.[15] The paper argued that a transformation of the JSDF would boost their effectiveness and reduce casualties in a future conflict. It would also improve combined operations with the United States. As the paper put it, the goal of a post-RMA military should be:

> Sharing real-time information among each unit of the Ground, Maritime, and Air-Self Defense Forces based on redundant and invulnerable information networks comprised of various sensors; securing interoperability between SDF and U.S. forces; and establishing a defense posture that could perform most efficiently with a minimum of reaction time, and could respond flexibly in accordance with rapidly changing situations.[16]

The JDA also raised the possibility of creating experimental units within the JSDF to explore new ways of war.

While Australian defense analysts see the U.S. approach to information-age warfare as too expensive, their Japanese counterparts see it as inapplicable to the political environment that governs Japan's defensive national security policy. Concepts and organizations designed to improve the speed, survivability, and lethality of power projection forces have limited utility to a nation whose constitution prohibits it from engaging in offensive operations. As Sugio Takashashi puts it, Japan's exploitation of the information revolution should focus upon defensive capabilities, such as leveraging IT to defend Japan against ballistic missiles. It should also improve interoperability with the United States.

China

China's efforts to exploit the emerging RMA arguably are the most focused of any of the states we examined. China's ongoing dispute with Taiwan, coupled with the prospect of U.S. intervention in a Taiwan Straits conflict, drives Beijing's modernization efforts. Chinese analysts have paid particular attention to IW as a relatively cheap way of countering the technological superiority of the United States. They have also outlined concepts to counter U.S. power projection forces, particularly carrier battle groups.[17]

In China, as elsewhere, there is no unanimity over the future of warfare. Indeed, You Ji identifies three schools of thought within the PLA regarding the RMA. He notes, however, that both Jiang Zemin and his successor, Hu Jintao, are RMA enthusiasts who have used their power to promote likeminded officers within the PLA. As a result, Ji predicts that support for the RMA will grow over time.

Chinese military authors have devoted considerable attention to future warfare concepts. The Chinese analyst Cheng Bingwen, for example, has coined the term "no-contact warfare" to discuss how long-range precision-strike systems may change the character and conduct of future conflicts. Drawing upon Operation *Allied Force*, NATO's bombing campaign over Serbia, he argues that air power will be the primary means of achieving victory in future conflicts; ground engagements will be shorter or nonexistent. He predicts that future conflicts will involve the use of long-range precision-strike weaponry to paralyze an enemy's command, telecommunications, and information systems; inflict heavy losses on military targets; and ultimately force it to yield. Success will depend upon identifying an adversary's vital points and weak links so that "hitting one point can paralyze a large area."[18] Indeed, Chinese theorists believe that they can employ IW to paralyze an adversary's leadership.[19]

China is attempting to harness science and technology to improve its armed forces. The Chinese military has accorded IT the highest priority in its modernization program. China is known to be developing doctrine and concepts for IW. It is studying the offensive employment of IW against foreign economic, logistics, and C4I systems and appears interested in researching methods to insert computer viruses into foreign computer networks.[20] Chinese authors have discussed forming a force dedicated to conducting information operations. Moreover, the Chinese armed forces have begun incorporating IW into their exercises.[21]

The Chinese military's pursuit of the RMA focuses on areas that the United States has abandoned, such as precision-guided conventional ballistic missiles, or is not interested in, such as methods to deny access to its sphere of interest. China is also pursuing so-called "assassin's mace" or "Trump Card" weapons that it hopes will give it an edge in a future conflict.[22]

China has begun fielding a new generation of precise ballistic missiles guided by signals from the Global Positioning System (GPS) satellite constellation. These missiles give Beijing the capability to target air defense installations, airfields, naval bases, C4I nodes, and logistics facilities. The People's Republic of China (PRC) has deployed the CSS-6 (DF-15 or M-9) SRBM, a road mobile missile that can deliver a 500-kg payload to a range of 600 km. It has developed the CSS-7 (M-11) SRBM, which has an estimated range of 300 km, and is developing an improved version of the CSS-7, the CSS-7 Mod 2, with greater range.[23] China has deployed approximately 450 SRBMs and this number is expected to increase by over 75 missiles per year over the next several years.[24] China has begun developing two land-attack cruise missile designs for conventional strike, programs that have a high development priority,[25] and has reportedly purchased Israeli *Harpy* unmanned combat air vehicles (UCAVs) for defense-suppression missions.[26]

China is also developing a range of advanced technologies. Beijing is improving its space capability, including intelligence-gathering, communication, and navigation satellites. It is also believed to be developing a ground-based laser anti-satellite systems. Still, Chinese modernization efforts face a series of constraints, including an aging capital stock, widespread corruption, and weak defense industries.

Taiwan

Taiwan's approach to the RMA is predicated upon developing the means to counter China, in particular to offset China's missile and IW capabilities. Taiwan is reportedly developing both a 1,000–2,000-km missile for use against deep targets and a 300-km system for striking targets along the China coast.[27] Taipei reportedly plans to deploy a land-attack cruise missile with a range exceeding 300 km.[28]

As Arthur Ding (chapter 8) and James Mulvenon (chapter 7) discuss, the Taiwanese concept of the RMA places heavy emphasis upon IW. Ding notes that developing IW has been a priority for Taiwan's armed forces since at least 1998. Taiwan, which manufactures nearly 80 percent of the computer chips in commercial use today, has all the basic capabilities needed to carry out offensive IW, particularly computer network attack and the introduction of malicious code. Within the Taiwanese military, the Communication Electronic and Information Bureau has been given responsibility for IW. In November 2000, the Taiwanese Ministry of National Defense announced that it had established an IW cell under the direct control of the General Staff Headquarters.[29] According to another report, the Taiwanese military has developed some 1,000 computer viruses that could be unleashed against China.[30] Taiwan has also conducted extensive research into computer network defense.[31] Still, Taiwan faces a number of economic and social constraints on its pursuit of new ways of war.

Singapore

Like the other militaries in this study, the Singapore Armed Forces (SAF) have debated the merits of the RMA. As Tim Huxley notes, senior civilian and military SAF leaders are fluent in the vocabulary of the RMA. *Defending Singapore in the 21st Century*, published in February 2000, explicitly discusses the desirability of the RMA for Singapore. The SAF journal *Pointer* has contained a number of articles debating the merits of information-age ways of war. In early 2003, the Ministry of Defense set up a Future Systems Directorate. In addition the SAF established a Center for Military Experimentation to test new operational concepts and provide modeling and simulation for the armed services.

At the strategic level, Singapore's approach to the emerging RMA is focused upon defending the small city-state against its larger neighbors.

Table 10.2 Niches being pursued by Asian militaries

	Australia	China	Japan	Singapore	Taiwan
Advanced C4ISR	X	X	X	X	X
Precision-strike	X	X		X	X
Information warfare	X	X	X	X	X
Space support	X[a]	X[b]	X		
Theater missile defense			X		X
Unmanned air warfare		X		X	

Notes:
[a] Through Australia's relationship with the United States.
[b] Including anti-satellite capabilities.

At the tactical level, the SAF seeks to acquire the ability to locate and destroy targets around the clock in joint operations. The SAF has emphasized IT, including C4ISR programs, offensive IW, and SIGINT.[32] The SAF has also expressed interest in UAVs, including a huge battle management drone.

As this discussion demonstrates, Asian militaries are pursuing a range of niche capabilities, including advanced C4ISR systems, long-range precision strike, IW, space support to terrestrial military operations, theater missile defense, and unmanned aerial warfare.[33] Table 10.2 summarizes the militaries that appear active in these niches.

Drivers of the Diffusion of the RMA in Asia

Each of the militaries that is discussed in this volume is pursuing the information revolution for its own purposes (see Table 10.3). There are two overarching—and overlapping—motives to exploit the RMA. First, Asian states are pursuing new approaches to combat in an effort to redress existing problems that defy a conventional solution. These include not only operational and strategic challenges, but also geographic and demographic predicaments. Second, they are attempting to exploit the benefits of new ways of war. These may include exploiting a comparative advantage in IT or increasing interoperability with the United States.

The RMA presents Australia with both challenges and opportunities. First, Australian analysts believe their country is losing its qualitative edge over regional neighbors. In the past, Australia could count upon enjoying a technological advantage over its adversaries. While the ADF will remain dominant in some areas, the proliferation of high-technology arms to the region is eroding Australia's edge. Second, the spread of long-range precision-strike weapons, overhead reconnaissance capabilities, and IW is reducing Australia's ability to rely upon its strategic depth. Taken together, these trends argue for exploring new approaches to combat. Third, because the ADF is so small, it cannot afford to take heavy casualties or lose many platforms. Information-age ways of war appear to offer the ability to strike at a distance while reducing the prospect of casualties.

Table 10.3 Drivers of the RMA in Asia

	Australia	Japan	Singapore	China	Taiwan
Threats and challenges	Loss of qualitative superiority to adversaries.	Demographics. Low casualty tolerance.	Lack of depth, manpower.	Need to coerce, defeat Taiwan.	Needto counter Chinese missile, IW programs.
	Reduced value of geographic depth.			Need to deter, defeat U.S. intervention.	Demographics.
	Low casualty tolerance.			China's enemies pursuring RMA.	
Opportunities	Increase interoperability with United States.	Spin-offs for private industry.			Exploit comparative advantage in IT.
	Exploit C4ISR to master Australia's geography.	Increase interoperability with United States.			Cost savings of IW.

Pursuing the RMA also holds advantages for Australia. Foremost among these is the need to maintain (or improve) interoperability with the United States. As the U.S. armed forces modernize their information infrastructure, the ADF must keep up or risk reduced relevance. Maintaining a high level of interoperability with the United States would enhance Australia's security, but it comes at a price. C4ISR modernization is expensive and resources expended on coalition capabilities may come at the expense of those required for unilateral defense or regional peacekeeping.

Exploiting the RMA would also help Australia master its geography. As Coral Bell has observed, "the Revolution in Military Affairs offers the most promising set of systems yet evolved to solve Australia's permanent strategic dilemma: how to defend a very large territory and a long and vulnerable coastline with forces which will always remain very small by global or regional standards."[34] And as Michael Evans notes, IT can provide the ADF with better surveillance of Australia's neighborhood and permit more efficient positioning and targeting of forces.

Japan, like Australia, faces a graying population. This, combined with a high sensitivity to casualties within Japanese society, is driving Tokyo to pursue ways of war that substitute technology for manpower. Pursuing the RMA holds a number of other benefits for Japan. First the development of IT will help not only the JSDF, but also Japanese industry. Moreover, as with

Australia, pursuing the RMA will increase interoperability with U.S. armed forces and enhance Japanese security.

Singapore seeks to exploit the emerging RMA to ensure the state's survival in a hostile regional environment. The SAF is interested in developing new ways of war to compensate for Singapore's lack of depth and shortage of manpower. Precision-guided munitions, electronic warfare, and information warfare offer Singapore the ability to deter and defend against its much larger neighbors. In addition, the SAF hopes to use the modernization of its information infrastructure as a means to enhance cooperation with regional powers, presumably including the United States.

China's primary motivation for pursuing the RMA is the need to coerce and (if necessary) defeat Taiwan while also deterring and (if necessary) defeating U.S. intervention in the conflict. China can more effectively strike Taiwan with precision-guided munitions and IW than with its air force, navy, and army. Moreover, the fact that those countries that are most vigorously pursuing the RMA are China's enemies also provides an impetus for Beijing to explore new ways of war.

The main driver of Taiwan's exploitation of the emerging RMA is its need to counter China's missile and IW programs. As Arthur Ding and James Mulvenon note, Taiwanese defense analysts perceive an imminent threat of attack by Chinese missiles and information operations. In addition, Taiwanese defense analysts see the RMA as a way to offset Taiwan's unfavorable demographic situation. Not only is Taiwan much smaller than mainland China, it also faces a decline in population growth and in the pool of available draftees.

Pursuing the RMA contains a number of potential benefits for Taiwan. Taiwan has a comparative advantage in IT. Indeed, it is one of the most "wired" countries in the world, has one of the world's most advanced information infrastructures, and ranks eighth in terms of the number of Internet users. Taiwan's pursuit of the RMA—and IW in particular—is seen as a way of exploiting these comparative advantages. The Taiwanese military also sees IW as a low-cost approach to combat.

CONSTRAINTS ON INNOVATION

Just as each of the militaries in this volume is pursuing the RMA for different reasons, so too do they face political, economic, military, and sociocultural constraints on innovation (see Table 10.4).

Australia faces political and military constraints on its exploitation of the RMA. First, any attempt to exploit new ways of war will have to contend with some stark fiscal realities. As Adam Cobb notes, the Australian defense budget is currently less than 2 percent of GNP and is unlikely to grow significantly. Because the Australian government has chosen to field a fairly robust force within these limits, much of the defense budget is spent on current operations. As a result, the ADF will face some stark choices in the short term.

Second, the ADF lacks a single, unifying conception of Australia's interests. Historically Labour governments have taken a narrow view, equating

Table 10.4 Constraints on the diffusion of the RMA in Asia

	Australia	Japan	Singapore	China	Taiwan
Political	Low defense spending. Lack of agreed upon national security vision.	Japanese constitution.		Dominance of People's War	Diplomatic isolation. Export restrictions. Declining defense budget.
Economic		Weak economy.		Limited technology base.	Strained resources. Insufficient indigenous technology base.
Military	Service culture "ANZAC Spirit" Block obsolescence of major systems.	Split within JSDF over RMA.	Military organizational culture.	Decision-making dominated by ground commanders, People's War	Military organizational culture.
Social/ cultural			Hierarchical culture.	Poor education and training.	Overly rigid society.

security with territorial defense, while the Liberal Party has been more expansive. Similarly, each service has its own concept of Australian security, that of the navy being most expansive and that of the army the narrowest. Third, Australia faces a number of organizational constraints on innovation. The Australian armed services, like their counterparts across the globe, resist changes that threaten cultural norms and disrupt hierarchies.

At a deeper level, the information age challenges the core beliefs of the ADF. Central to the ADF is the concept of the "ANZAC Spirit," which places a premium on the skill and initiative of the individual soldier. This culture could be of decreasing importance in the future, and might even become dysfunctional.

Despite growing interest in transformation in Japan, such efforts face substantial political, legal, and social barriers. For example, Japan's constitution prevents it from acquiring offensive weapons. Moreover, as Sugio Takahashi (chapter 4) points out, supporters of exploiting the RMA form a minority within the JSDF. Japan's sluggish economy also poses a major brake on efforts to modernize the Japanese armed forces. And while Tokyo is a world leader in IT, Japanese weapons development has at times been glacially slow. Any attempt at transformation will thus face daunting challenges.

While Singapore appears to be pursuing IT quite aggressively, it nonetheless faces military and socio-cultural barriers of its own. Although the SAF is perhaps the best-equipped and trained military in the region, its organizational culture, which values hierarchy and discipline over innovation, is a weakness. Similarly, while the SAF has embraced new technologies and weapon systems, it has shown less willingness to innovate doctrinally or organizationally.

China also faces constraints upon its exploitation of the RMA. Andrew Yang argues that China's scientific and technical infrastructure represents a significant weakness. In his view, the Chinese government provides insufficient funding to high-technology projects. Moreover, the PLA suffers from a lack of well-trained scientists and a backward research and development infrastructure.

Less concretely, the PLA faces a number of political and cultural barriers to the adoption of new ways of war. These include the fact that many in leading positions in the PLA have a vested interest in the *status quo*. In particular, Andrew Yang and You Ji point out that officers who equate Chinese security with homeland defense currently dominate China's national security decision-making. Moreover, the fact that ground force commanders have a near monopoly on power within the PLA limits the voice of the navy and air force in decision-making.

Taiwan faces a range of constraints in its efforts to exploit the RMA. Taiwan's diplomatic isolation and restrictions on the export of weapons to Taiwan both limit Taipei's ability to exploit the information revolution. A declining defense budget, strained scientific and technical resources, and a limited indigenous technology base compound the problem.

Like the other states in this study, perhaps the most intractable problems that Taiwan faces are cultural. Both Arthur Ding and James Mulvenon note that efforts to develop information warfare have encountered opposition from traditionalists within Taiwan's armed forces. Overly rigid discipline and lack of innovation within the military has been another problem. As a result, Taiwan's development of IW has emphasized technology over doctrine and organizational coordination.

CONCLUSION

Australia appears to be the farthest along at institutionalizing new ways of war, followed by China. Singapore has also made great strides toward fielding an information-age military. Japan and Taiwan's efforts have been much more limited. The history of past transformations shows that states that take an early lead in exploiting new ways of war do not always sustain that lead. While Britain was an early leader in exploiting the tank in the 1920s, it was Germany that ultimately developed combined-arms armored warfare. Similarly, today's leaders may fall behind, to be replaced by militaries that are currently followers.

This project has revealed a diversity of attitudes toward the emerging RMA in Asia. States in the region are pursuing new ways of war in response to a range of drivers and constraints. If the history of past revolutions of warfare is a guide, the tendency toward divergence in military technology, doctrine, and organization is likely to continue until there is a convincing battlefield demonstration of some new way of war. It is only after such a demonstration that we would expect a "model" to spread throughout the region.

NOTES

1. In this respect I disagree with Chris Demchak, "Complexity and a Midrange Theory of Networked Militaries," in Theo Farrell and Terry Terriff, eds., The *Sources of Military Change: Culture, Politics, Technology* (Boulder, CO: Lynne Rienner, 2002).

2. On the diffusion of German combined-arms armored warfare methods and organizations, see Thomas G. Mahnken, "Beyond *Blitzkrieg*: Allied Responses to Combined-Arms Armored Warfare During World War II" in Emily O. Goldman and Leslie C. Eliason, eds., The *Diffusion of Military Technology and Ideas* (Stanford: Stanford University Press, 2003).

3. It is worth noting, for example, that all the "strategists" cited in the chapter devoted to "strategists and the revolution in military affairs" in the Australian Defence Studies Centre's textbook on strategy are Americans. See Steven Metz, "Strategists and the Revolution in Military Affairs," in Hugh Smith, ed., The *Strategists* (Canberra: Australian Defence Studies Centre, 2001) chapter 9.

4. Australian Defence Force, *Future Warfighting Concept*, ADDP-D.3 (Canberra: Department of Defence, 2003).

5. Ibid., pp. 11–13.

6. See, for example, the proceedings of the first Australian RMA conference in Keith Thomas, ed., The *Revolution in Military Affairs: Warfare in the Information Age* (Canberra: Australian Defence Studies Centre, 1997).

7. Indeed, Liddell-Hart and Fuller had greater influence abroad than in Britain. See Azar Gat, "British Influence and the Evolution of the Panzer Arm: Myth or Reality? Part I," *War in History*, 4:2 (April 1997) p. 160 and *passim*.

8. See, for example, Lieutenant Colonel Greg de Somer, The *Implications of the United States Army's Army-After-Next Concepts for the Australian Army*, Land Warfare Studies Centre Working Paper no. 104 (Canberra: Land Warfare Studies Centre, June 1999).

9. On India's approach to the emerging RMA, see Thomas G. Mahnken and Timothy D. Hoyt, "Indian Views of the Emerging Revolution in Military Affairs," *National Security Studies Quarterly*, 6:3 (Summer 2000).

10. *Annual Report on the Military Power of the People's Republic of China*, Report to Congress Pursuant to the FY2000 National Defense Authorization Act (Washington, D.C.: Department of Defense, July 28, 2003) p. 19.

11. Commonwealth of Australia, *Australia's Strategic Policy 1997* (Canberra: Directorate of Publishing and Visual Communications, 1997) p. 56.

12. Michael Sinclair-Jones, "JORN Assures Early Warning for Australia," *Defence Systems Daily*, February 29, 2000 at http://defence-data.com/features/fpage37.htm (accessed July 5, 2001).

13. Admiral Robert J. Natter, "Meeting the Need for Speed," *Proceedings*, 128, no. 6 (June 2002) pp. 65–67.

14. Michael D. Swaine, Rachel M. Swanger, and Takashi Kawakami, *Japan and Ballistic Missile Defense* (Santa Monica: The RAND Corporation, 2001).

15. Office of Strategic Studies, Defense Policy Division, Defense Policy Bureau, Japan Defense Agency, *Info-RMA: Study on Info-RMA and the Future of the Self-Defense Forces* (Tokyo: JDA, December 2000) (http://www.jda.go.jp/e/pab/rma/rma_e.pdf).

16. Ibid., p. 9.1.

17. See, for example, Zhou Yi, "Aircraft Carrier Face Five Major Assassins," *Junshi Wenzhai*, March 1, 2002, pp. 4–6.

18. Cheng Bingwen, "Countermeasures and Thoughts for Fighting 'No-Contact Warfare'—on the Need to Refocus Our Preparations for Military Struggles," *Jiefangjun Bao* (Internet Version) in Chinese, October 4, 1999, p. 3.

19. *Annual Report on the Military Power of the People's Republic of China, Pursuant to the FY2000 National Defense Authorization Act*, at http://www.defenselink.mil/news/Jun2000/china06222000.htm (accessed July 2000).

20. *The Security Situation in the Taiwan Strait*, Report to Congress Pursuant to the FY99 Appropriations Bill, 12.

21. Wen T'ao, "PLA Bent on Seizing 'Information Control'," *Ching Pao*, June 1, 2002, pp. 44–46.

22. *Annual Report on the Military Power of the People's Republic of China*, Report to Congress Pursuant to the FY2000 National Defense Authorization Act (Washington, D.C.: Department of Defense, July 28, 2003) p. 21.

23. *The Security Situation in the Taiwan Strait*, p. 5.

24. *Annual Report on the Military Power of the People's Republic of China*, Report to Congress Pursuant to the FY2000 National Defense Authorization Act (Washington, D.C.: Department of Defense, July 28, 2003) p. 22.

25. *Annual Report on the Military Power of the People's Republic of China, Pursuant to the FY2000 National Defense Authorization Act*, at http://www.defenselink.mil/news/Jun2000/china06222000.htm (accessed July 2000).

26. Bill Gertz, "China Deploys Drones from Israel," *Washington Times*, July 2, 2002, p. 1.

27. Robert Karniol, "Taiwan's Survival Strategy," *Jane's Defence Weekly*, September 13, 2000. See also Barbara Opall-Rome, "Support Mounts in Taiwan for Ballistic Missiles," *Defense News*, April 26, 1999 and Frank Umbach, "World Gets Wise to Pyongyang's Nuclear Blackmail," *Jane's Intelligence Review*, October 1, 1999.

28. Karniol, "Taiwan's Survival Strategy."

29. "Taiwan Military to Form Cyber Warfare Unit," *Defense Systems Daily* (web based) (accessed November 30, 2000).

30. "Taiwan Computer Viruses to Defend Against PRC Attack," AFP, January 9, 2000.

31. See Chung-Yang Jih Pao (internet version) November 22, 1999.

32. On Singapore's SIGINT capabilities, see Desmond Ball, *Developments in Signals Intelligence and Electronic Warfare in Southeast Asia*, Strategic and Defence Studies Centre Working Paper no. 290 (Canberra: Australian National University, December 1995) pp. 16–18.

33. Other Asian states interested in the RMA include South Korea and India. On South Korea, see Jiyul Kim and Michael J. Finnegan, "The Republic of Korea Approaches the Future," *Joint Force Quarterly*, no. 30 (Spring 2002) pp. 33–40. On India, see Thomas G. Mahnken and Timothy D. Hoyt, "Indian Views of the Emerging Revolution in Military Affairs," *National Security Studies Quarterly*, 6: 3 (Summer 2000).

34. Coral Bell, "Security Regionalisation and the Future of the Australian Defence Forces," *Australian Defence Force Journal*, no. 143 (July/August 2000) p. 21.

INDEX

downsizing of forces
  People's Liberation Army (PLA), 109–10, 117, 127, 134, 135
  Taiwan, 149, 173
drones. *See* unmanned aerial vehicles (UAVs)
DSTO. *See* Defense Science and Technology Organization
dual-use technologies, 190
Dudley, Jonathan, 74
Dunn, Peter, 25–6

early warning networks
  China and, 105
  Singapore and, 198
  Taiwan and, 148, 153–60, 169, 170
  See also radar systems
East Timor, Australian involvement in, 33, 39, 67, 70, 74
economic factors, 5, 9–11
  See also specific countries
EDOP ("exclusively defense oriented policy") of Japan, 84, 85–6
education and training
  advanced technologies and society's level in, 12
  Australia with U.S., 211
  China and training of forces, 116, 127
  Japan with U.S., 211
  regime preservation and military training, 9
  Singapore: general population, 187–8; military forces, 199
  Taiwan: information-savvy population of, 140–1, 172, 179, 220; training of forces, 174, 176, 179
effects-based operations
  Australian focus on, 43, 64–5, 210
  U.S. model of, 210
efficiency principle of RMA, 88
Eisenstadt, Michael J., 8, 12
electronic warfare. *See* missiles and electronic warfare systems
elite forces, China's need to develop, 103
"Enabling Multidimensional Maneuver" (draft ADF discussion paper), 47, 212
England. *See* Britain

"exclusively defense oriented policy" (EDOP) of Japan, 84, 85–6
experimentation as phase of innovation, 4

Fahey, John, 39
ferry, high-speed, 214
Force 2020 (Australia), 42, 64
Foster, William, 8
Foundations of Australian Military Doctrine, 37
France in collaborative R&D program with Singapore, 191
Future Combat System of U.S. Army, 193
Future Warfighting Concept (Australia), 42, 43–7, 64, 65
Fu Wei-ku, 169

Garran, Robert, 39
Gartska, John J., 45
Ge Zhenfeng, 108
Gill, Bates, 10
global positioning systems (GPS), 27, 130, 210, 216
global strategic environment, 2–3, 59–62
GLONASS (Global Navigation Satellite system), 130
Goldstone, Jack A., 13
Goldwater-Nichols Act of 1986, 82
Goodman, Seymour E., 8
GPS. *See* global positioning systems
Guidelines for Comprehensive Measures for Information Technology Revolution of the JDA/SDF (Japan), 88–9
"Guidelines for Defense Cooperation between Japan and the U.S.", 85
Gulf War (1991), 62–3, 67, 81, 89, 98, 139
Guo Boxiong, 108

hackers. *See* cyber attacks and protection
Hall, John A., 12
Hart, Liddell, 45
Hawke, Allen, 68–9, 71
helicopters, 195
Henley, Lonnie, 10
high-speed ferry, use of, 214
Hill, Robert, 24, 46, 48

230

INDEX

Hofstede, Geert, 13
homeland security
of Australia, 41–2, 48, 59, 72–3, 218,
220–1
of China, 100–1
of Singapore, 203
as spur to innovation, 8
Howard, John, 26, 28, 39, 40
Hoyt, Timothy D., 7
Huang, Alexander, 144, 146
Hu Changfa, 111, 112
Hu Jintao, 103, 216
human capital theory, 12
Hung, Alan, 150

Ikenberry, G. John, 12
implementation as phase of
innovation, 4
Improved Mobile Subscriber Equipment
(IMSE), 171
InCat vessels, 74
India, 11, 99, 197, 212
indirect approach strategy, 45
Indonesia, 61, 202–3
*Info-RMA: Study on Info-RMA and the
Future of the Self-Defense Forces*
(Japan), 88
information networks, development
of, 82
information sharing, importance of, 89
information technology
advanced information systems of
Japan, 88–9
Artillery Tactical Command and
Control System (ATCCS), 196
Australia and, 33, 35–7
China and, 102–3, 104–5, 109, 111,
127, 216
commercial-off-the-shelf (COTS)
hardware and software, 189, 196
dissemination issues, 13–14
INSIGHT system, 196
Singapore and, 186, 189, 196, 200
software development and training
issues, 7
Taiwan and, 140–1, 156–7, 160,
176–7, 179
See also cyber attacks and protection
*Information Technology Revolution of
JDA/SDF* (Japan), 88

information warfare (IW) systems
ambiguous use of term, 210
Asian militaries pursuing, 218
Australia's vulnerability to, 218
China's development of, 98, 113,
129, 215, 216
Japanese development of, 88
as part of U.S. RMA, 82
spread of, 210–11
Taiwanese development of, 154–60,
171–2, 178, 210, 217
U.S. use of, 210
See also cyber attacks and protection;
missiles and electronic warfare
systems
innovation
acceleration of efforts abroad, 3
constraints on, 220–2
defined, 1
distinguished from diffusion, 5
enablers and inhibitors of, 7–16
institutional resistance to, 6
phases of, 4
See also specific countries
INSIGHT, 196
Institutes of Systems Science,
Information Technology and High
Performance Computing, 190
Internet. *See* cyber attacks and
protection; information technology
interoperability
allies seeking to optimize, 5
Australia and U.S., 23, 26, 32, 36,
37, 219
Japan and U.S., 89–90, 92, 215, 220
Singapore and U.S., 204
Iraq, 23, 101
See also Gulf War (1991); Operation
Iraqi Freedom
Israel
R&D cooperation with Singapore,
191, 194, 197, 198
satellite, use by Taiwan, 170
IW. *See* information warfare (IW)
systems

Japan, 81–95, 214–15
adoption of RMA, 2, 82–4, 222
Air Self-Defense Forces (JASDF),
85, 89

GPSR Compliance
The European Union's (EU) General Product Safety Regulation (GPSR) is a set
of rules that requires consumer products to be safe and our obligations to
ensure this.

If you have any concerns about our products, you can contact us on

ProductSafety@springernature.com

In case Publisher is established outside the EU, the EU authorized
representative is:

Springer Nature Customer Service Center GmbH
Europaplatz 3
69115 Heidelberg, Germany